CliffsNotes® Basic Math & Pre-Algebra Quick Review

2nd Edition

By Jerry Bobrow, Ph.D.

Revised by Ed Kohn, M.S.

Wiley Publishing, Inc.

About the Author

Jerry Bobrow, Ph.D., was an award-winning teacher and educator, and his company BTPS Testing is a national authority in the field of test preparation. BTPS Testing has been administering test preparation programs for most California State Universities for the past 34 years. Dr. Bobrow and his faculty have authored more than 30 national best-selling test preparation books including Cliffs Preparation Guides for the GRE, GMAT, CSET, SAT, CBEST, ACT, and PPST. Each year the faculty of BTPS Testing lectures to thousands of students on preparing for these important exams.

Publisher's Acknowledgments

Editorial

Acquisitions Editor: Greg Tubach
Project Editor: Kelly D. Henthorne
Technical Editors: David Herzog and Mary Jane Sterling
Contributing Author: Ron Podrasky, M.A.

Composition

Indexer: BIM Indexing & Proofreading Services
Proofreader: Henry Lazarek
Wiley Publishing, Inc. Composition Services

CliffsNotes® Basic Math & Pre-Algebra Quick Review, 2nd Edition
Published by:
Wiley Publishing, Inc.
111 River Street
Hoboken, NJ 07030-5774
www.wiley.com

Published by Wiley, Hoboken, NJ
Published simultaneously in Canada

Library of Congress Control Number: 2011922789
ISBN: 978-0-470-88040-1 (pbk)
ISBN: 978-1-118-01761-6 (ebk)

Printed in the United States of America
10 9 8 7 6 5 4 3 2 1

WILEY

Table of Contents

INTRODUCTION

C*liffsNotes Basic Math & Pre-Algebra Quick Review, 2nd Edition* is designed to give a clear, concise, easy-to-use review of arithmetic and pre-algebra. Introducing each topic, defining key terms, and carefully walking through each sample problem type in a step-by-step manner gives the student insight and understanding of the important basics of mathematics.

The prerequisite to get the most out of this book is a familiarity with the basic concepts of arithmetic—working with fractions, decimals, and percents. This book starts with the more basic concepts of arithmetic and builds upon these concepts as you work through pre-algebra.

Why You Need This Book

Can you answer "yes" to any of these questions?

- Do you need to review the fundamentals of math and pre-algebra quickly?
- Do you need a course supplement to pre-algebra?
- Do you need a concise, comprehensive reference for arithmetic and pre-algebra?

If so, then *CliffsNotes Basic Math & Pre-Algebra Quick Review* is for you!

How to Use This Book

You can use this book in any way that fits your personal style for study and review—you decide what works best with your needs. You can either read the book from cover to cover or just look for the information you want and put it back on the shelf for later. Here are just a few ways you can search for topics:

- Look for areas of interest in the book's table of contents or use the index to find specific topics.
- Flip through the book, looking for subject areas at the top of each page.

- Get a glimpse of what you'll gain from a chapter by reading through the "Chapter Check-In" at the beginning of each chapter.
- Use the "Chapter Check-Out" at the end of each chapter to gauge your grasp of the important information you need to know.
- Test your knowledge more completely in the "Review Questions" and look for additional sources of information in the "Resource Center."
- Use the glossary to find key terms fast. This book defines new terms and concepts where they first appear in the chapter. If a word is boldfaced, you can find a more complete definition in the book's glossary.
- Or flip through the book until you find what you're looking for—we organized this book to gradually build on key concepts.

Hundreds of Practice Questions Online!

Go to CliffsNotes.com for hundreds of additional basic math and pre-algebra practice questions to help prepare you for your next quiz or test. The questions are organized by this book's chapter sections, so it's easy to use the book and then quiz yourself online to make sure you know the subject. Go to www.cliffsnotes.com to test yourself anytime and find other free homework help.

Chapter 1
PRELIMINARIES

Chapter Check-In

- ❑ Groups of numbers
- ❑ Common math symbols
- ❑ Properties of addition and multiplication
- ❑ Grouping symbols
- ❑ Order of operations

Before you begin reviewing basic math or pre-algebra, you need to feel comfortable with some terms, symbols, and operations.

Groups of Numbers

In doing basic math, you work with many different groups of numbers. The more you know about these groups, the easier they are to understand and work with.

- Natural or counting numbers: 1, 2, 3, 4, . . .
- Whole numbers: 0, 1, 2, 3, 4, . . .
- Integers: . . . –3, –2, –1, 0, 1, 2, 3, . . .
- Negative integers: . . . –3, –2, –1
- Positive integers: 1, 2, 3, . . . (the natural numbers)

  Note: Zero is neither positive nor negative. It is neutral.

- Odd numbers; integers not divisible by 2:

  $\ldots -5, -3, -1, 1, 3, 5, \ldots$

- Even numbers: Integers divisible by 2:

  $\ldots -6, -4, -2, 2, 4, 6, \ldots$

- Rational numbers: Fractions, such as $\frac{3}{5}$ or $\frac{7}{8}$. All **integers** are **rational numbers;** for example, the number 5 may be written as $\frac{5}{1}$. All rational numbers can be written as fractions $\frac{a}{b}$, with a being an integer and b being a natural number. Both terminating decimals (such as 0.5) and repeating decimals (such as 0.333. . .) are also rational numbers because they can be written as fractions in this form.
- **Irrational numbers:** Numbers that *cannot* be written as fractions $\frac{a}{b}$, with a being an integer and b being a natural number. $\sqrt{3}$ and π (the Greek letter pi) are examples of irrational numbers.

Ways to Show Multiplication and Division

Some operations can be written in a variety of forms. You should become familiar with the different forms. Following are several ways to show multiplication:

- Multiplication sign: $4 \times 3 = 12$
- Multiplication dot: $4 \cdot 3 = 12$
- Two sets of parentheses: $(4)(3) = 12$
- One set of parentheses: $4(3) = 12$ or $(4)3 = 12$
- A number next to a variable (letter): $3a$ means 3 times a.
- Two variables (letters) next to each other: ab means a times b.

There are several ways to show division:

- **Division sign:** $10 \div 2 = 5$
- **Fraction bar:** $\frac{10}{2} = 5$ or $10/2 = 5$

Multiplying and Dividing Using Zero

Zero times any number equals zero.

$$0 \times 2 = 0$$

$$8 \times 2 \times 3 \times 6 \times 0 = 0$$

Likewise, zero divided by any nonzero number is zero.

$$0 \div 3 = 0$$

$$\frac{0}{7} = 0$$

Note: Dividing by zero is "undefined" and is not permitted. For example, $\frac{4}{0}$ and $\frac{0}{0}$ are not permitted because there is no logical answer. The answer is not zero.

Common Math Symbols

The following symbols are commonly used in basic math and algebra. Be sure to know what each symbol represents.

$=$ is equal to

$\neq$ is not equal to

$>$ is greater than

$<$ is less than

$\geq$ is greater than or equal to (also written $\geqq$)

$\leq$ is less than or equal to (also written $\leqq$)

$\not>$ is not greater than

$\not<$ is not less than

$\not\geq$ is not greater than or equal to

$\not\leq$ is not less than or equal to

$\approx$ is approximately equal to (also written $\doteq$)

Properties of Basic Mathematical Operations

Some mathematical operations have properties that can make them easier to work with and actually can save you time.

Some properties (axioms) of addition

- **Closure** is when all answers fall into the original set. If you add two even numbers, the answer is still an even number ($2 + 4 = 6$); therefore, the set of even numbers *is closed* under addition (has closure). If you add two odd numbers, the answer is not an odd number ($3 + 5 = 8$); therefore, the set of odd numbers *is not closed* under addition (no closure).

- **Commutative** means that the *order* does not make any difference in the operation's result.

  $$2 + 1 = 1 + 2$$

  $$a + b = b + a$$

  Note: Commutative does *not* hold for subtraction.

  $$3 - 2 \neq 2 - 3$$

  $$1 \neq -1$$

  $$a - b \neq b - a$$

- **Associative** means that the *grouping* does not make any difference in the operation's result.

  $$(2 + 3) + 4 = 2 + (3 + 4)$$

  $$(a + b) + c = a + (b + c)$$

  The grouping has changed (parentheses moved), but the sides are still equal.

  Note: Associative does *not* hold for subtraction.

  $$4 - (2 - 1) \neq (4 - 2) - 1$$

  $$4 - 1 \neq 2 - 1$$

  $$3 \neq 1$$

  $$a - (b - c) \neq (a - b) - c$$

- The **identity element** for addition is 0. Any number added to 0 gives you the original number.

$$5 + 0 = 5$$

$$a + 0 = a$$

- The **additive inverse** is the opposite (negative) of the number. Any number plus its additive inverse equals 0 (the identity).

$3 + (-3) = 0$; therefore, 3 and –3 are additive inverses.

$-4 + 4 = 0$; therefore, –4 and 4 are additive inverses.

$a + (-a) = 0$; therefore, a and $-a$ are additive inverses.

Some properties (axioms) of multiplication

- **Closure** is when all answers fall into the original set. If you multiply two even numbers, the answer is still an even number ($2 \times 4 = 8$); therefore, the set of even numbers is *closed* under multiplication (has closure). If you multiply two odd numbers, the answer is an odd number ($3 \times 5 = 15$); therefore, the set of odd numbers *is closed* under multiplication (has closure).

- **Commutative** means that the *order* does not make any difference in the operation's result.

$$4 \times 3 = 3 \times 4$$

$$a \times b = b \times a$$

Note: Commutative does *not* hold for division.

$$12 \div 4 \neq 4 \div 12$$

$$\frac{12}{4} \neq \frac{4}{12}$$

$$3 \neq \frac{1}{3}$$

$$a \div b \neq b \div a$$

- **Associative** means that the *grouping* does not make any difference in the operation's result.

$$(2 \times 3) \times 4 = 2 \times (3 \times 4)$$

$$(a \times b) \times c = a \times (b \times c)$$

The grouping has changed (parentheses moved), but the sides are still equal.

Note: Associative does *not* hold for division.

$$(8 \div 4) \div 2 \neq 8 \div (4 \div 2)$$

$$2 \div 2 \neq 8 \div 2$$

$$1 \neq 4$$

$$(a \div b) \div c \neq a \div (b \div c)$$

- The **identity element** for multiplication is 1. Any number multiplied by 1 gives the original number.

$$5 \times 1 = 5$$

$$a \times 1 = a$$

- The **multiplicative inverse** is the **reciprocal** of the number. Any nonzero number multiplied by its reciprocal equals 1.

$2 \times \frac{1}{2} = 1$; therefore, 2 and $\frac{1}{2}$ are multiplicative inverses, or reciprocals.

$a \times \frac{1}{a} = 1$; therefore, a and $\frac{1}{a}$ are multiplicative inverses, or reciprocals (provided $a \neq 0$).

A property of two operations

The **distributive property** is the process of distributing, using multiplication, the number on the outside of the parentheses to each term on the inside. The terms within the parentheses are separated by either addition or subtraction.

$$2(3 + 4) = 2(3) + 2(4)$$

$$a(b + c) = a(b) + a(c)$$

$$4(7 - 3) = 4(7) - 4(3)$$

$$a(b - c) = a(b) - a(c)$$

Note: You cannot use the distributive property with only one operation.

$$3(4 \times 5 \times 6) \neq 3(4) \times 3(5) \times 3(6)$$
$$3(120) \neq 12 \times 15 \times 18$$
$$360 \neq 3240$$
$$a(bcd) \neq a(b) \times a(c) \times a(d) \text{ or } (ab)(ac)(ad)$$

Grouping Symbols and Order of Operations

Three common types of grouping symbols—parentheses ( ), brackets [ ], and braces { }—are used to group *numbers* or *variables* (letters). The most commonly used grouping symbols are **parentheses.** Operations inside parentheses can be done before any other operations in order to simplify the problem.

Example 1: Simplify 4(3 + 5).

$$\begin{aligned} 4(3 + 5) &= 4(8) \\ &= 32 \end{aligned}$$

Example 2: Simplify (2 + 5)(3 + 4).

$$\begin{aligned} (2 + 5)(3 + 4) &= (7)(7) \\ &= 49 \end{aligned}$$

Brackets and **braces** are less commonly used grouping symbols and should be used after parentheses. Parentheses are to be used first, then brackets, and then braces: { [( )] }. Sometimes, instead of brackets or braces, you will see the use of larger parentheses.

Example 3: Simplify ((2+3) × 4) + 1.

$$\begin{aligned} ((2+3) \times 4) + 1 &= ((5) \times 4) + 1 \\ &= (20) + 1 \\ &= 21 \end{aligned}$$

An expression using all three grouping symbols looks like this:

$$2\{1 + [4(2+1) + 3]\}$$

Example 4: Simplify 2{1 + [4(2+1) + 3]}.

Notice that you work from the inside out.

$$\begin{aligned} 2\{1 + [4(2+1) + 3]\} &= 2\{1 + [4(3) + 3]\} \\ &= 2\{1 + [12 + 3]\} \\ &= 2\{1 + [15]\} \\ &= 2\{16\} \\ &= 32 \end{aligned}$$

Order of Operations

If multiplication, division, powers, addition, parentheses, and so on, are all contained in one problem, the **order of operations** is as follows.

1. parentheses
2. exponents
3. multiplication } whichever comes first left to right
4. division
5. addition } whichever comes first left to right
6. subtraction

An easy way to remember the order of operations is **P**lease **E**xcuse **M**y **D**ear **A**unt **S**ally (**P**arentheses, **E**xponents, **M**ultiplication/**D**ivision, **A**ddition/**S**ubtraction).

Example 5: Simplify each of the following.

(a) $16 + 4 \times 3$ **(b)** $10 - 3 \times 6 + 10^2 + (6 + 1) \times 4$

(a) First, the multiplication,

$$16 + 4 \times 3 = 16 + 12$$

Then, the addition,

$$16 + 12 = 28$$

(b) First, the parentheses,

$$10 - 3 \times 6 + 10^2 + (6 + 1) \times 4 = 10 - 3 \times 6 + 10^2 + (7) \times 4$$

Then, the exponents,

$$10^2 = 10 \times 10 = 100,$$

so $10 - 3 \times 6 + 10^2 + (7) \times 4 = 10 - 3 \times 6 + 100 + (7) \times 4$

Then, multiplication,

$$10 - 3 \times 6 + 100 + (7) \times 4 = 10 - 18 + 100 + 28$$

Then, addition and subtraction left to right,

$$\begin{aligned} 10 - 18 + 100 + 28 &= -8 + 100 + 28 \\ &= 92 + 28 \\ &= 120 \end{aligned}$$

Chapter Check-Out

1. Which of the following are integers? $-4, 6, \frac{1}{2}, 4.5$
2. Which of the following are rational numbers? $3, 4.9, -6, \pi$
3. True or False: > means "is greater than."
4. True or False: $3 + 5 = 5 + 3$ is an example of the commutative property of addition.
5. The identity element in multiplication is ______.
6. What is the additive inverse of -3?
7. True or False: $3(4 + 5) = 3(4) + 4(5)$.
8. Simplify: $3[4 + 6(2 + 5) - 3]$.

Answers: 1. –4, 6 **2.** 3, 4.9, –6 **3.** True **4.** True **5.** 1 **6.** +3 or 3 **7.** False **8.** 129

Chapter 2
WHOLE NUMBERS

Chapter Check-In

- ❑ Place value
- ❑ Expanded notation
- ❑ Rounding off
- ❑ Estimating answers
- ❑ Divisibility rules
- ❑ Factors
- ❑ Primes
- ❑ Factor trees

Place Value

Our number system is a **place value system**; that is, each place is assigned a different value. For instance, in the number 675, the 6 is in the hundreds place, the 7 is in the tens place, and the 5 is in the ones place. Because the number system is based on powers of 10 ($10^0 = 1$, $10^1 = 10$, $10^2 = 10 \times 10 = 100$, $10^3 = 10 \times 10 \times 10 = 1{,}000$, and so on), each place is a progressive power of ten, as shown in Table 2-1.

Table 2-1 Place value system

billions	hundred millions	ten millions	millions	hundred thousands	ten thousands	thousands	hundreds	tens	ones
1,000,000,000	100,000,000	10,000,000	1,000,000	100,000	10,000	1,000	100	10	1
10^9	10^8	10^7	10^6	10^5	10^4	10^3	10^2	10^1	10^0
							6	7	5

Notice how the number 675 fits into the place value grid.

Expanded notation

Sometimes, numbers are written in **expanded notation** to point out the place value of each digit.

Example 1: Write 523 in expanded notation.

$$523 = 500 + 20 + 3$$

$$= (5 \times 100) + (2 \times 10) + (3 \times 1)$$

$$= (5 \times 10^2) + (2 \times 10^1) + (3 \times 10^0)$$

These last two are the more common forms of expanded notation: one without exponents, one with exponents. Notice that in these the digit is multiplied times its place value: 1's, 10's, 100's, and so on.

Example 2: Write 28,462 in expanded notation.

$$28{,}462 = 20{,}000 + 8{,}000 + 400 + 60 + 2$$

$$= (2 \times 10{,}000) + (8 \times 1{,}000) + (4 \times 100) + (6 \times 10) + (2 \times 1)$$

$$= (2 \times 10^4) + (8 \times 10^3) + (4 \times 10^2) + (6 \times 10^1) + (2 \times 10^0)$$

Rounding off

To **round off** any number,

1. Underline the place value to which you're rounding off.
2. Look to the immediate right (one place) of your underlined place value.
3. Identify the number (the one to the right). If it is 5 or higher, round your underlined place value up 1 and change all the other numbers to its right to zeros. If the number (the one to the right) is 4 or less, leave your underlined place value as it is and change all the other numbers to its right to zeros.

Example 3: Round off 345,678 to the nearest thousand.

The number $34\underline{5},678$ is rounded up to 346,000.

Example 4: Round off 724,591 to the nearest ten thousand.

The number $7\underline{2}4,591$ is rounded down to 720,000.

Numbers that have been rounded off are called **rounded numbers**.

Estimating Sums, Differences, Products, and Quotients

Knowing how to approximate or estimate not only saves you time but can also help you check your answer to see whether it is reasonable.

Estimating sums

Use rounded numbers to estimate sums.

Example 5: Give an estimate for the sum 3,741 + 5,021 rounded to the nearest thousand.

$$\begin{array}{ll} & 3{,}741 + 5{,}021 \\ & \ \ \downarrow \qquad \downarrow \\ & 4{,}000 + 5{,}000 = 9{,}000 \\ \text{So,} & 3{,}741 + 5{,}021 \approx 9{,}000 \end{array}$$

Note: The symbol ≈ means *is approximately equal to.*

Estimating differences

Use rounded numbers to estimate differences.

Example 6: Give an estimate for the difference 317,753 – 115,522 rounded to the nearest hundred thousand.

$$\begin{array}{l} 317{,}753 - 115{,}522 \\ \quad\downarrow \qquad\quad \downarrow \\ 300{,}000 - 100{,}000 = 200{,}000 \end{array}$$

$$\text{So, } 317{,}753 - 115{,}522 \approx 200{,}000$$

Estimating products

Use rounded numbers to estimate products.

Example 7: Estimate the product of 722 × 489 by rounding to the nearest hundred.

$$\begin{array}{l} 722 \times 489 \\ \;\downarrow \quad\; \downarrow \\ 700 \times 500 = 350{,}000 \end{array}$$

$$\text{So, } 722 \times 489 \approx 350{,}000$$

If both multipliers end in 50 or are halfway numbers, then rounding one number up and one number down will give you a better estimate of the product.

Example 8: Estimate the product of 650 × 350 by rounding to the nearest hundred.

$$\begin{array}{ll} & 650 \times 350 \\ & \;\downarrow \quad\; \downarrow \\ \text{Round one number up and one down} & 700 \times 300 = 210{,}000 \end{array}$$

$$\text{So, } 650 \times 350 \approx 210{,}000$$

You can also round the first number down and the second number up and get this estimate:

$$\begin{array}{l} 650 \times 350 \\ \;\downarrow \quad\; \downarrow \\ 600 \times 400 = 240{,}000 \end{array}$$

$$\text{So, } 650 \times 350 \approx 240{,}000$$

In either case, your approximation is closer than it will be if you round both numbers up, which is the standard rule.

Estimating quotients

Use rounded numbers to estimate quotients.

Example 9: Estimate the quotient of 891 ÷ 288 by rounding to the nearest hundred.

$$\begin{aligned} 891 \div 288& \\ \downarrow \quad \downarrow& \\ 900 \div 300 = 3& \\ \text{So } 891 \div 288 \approx 3& \end{aligned}$$

Divisibility Rules

The following set of rules can help you save time in trying to check the divisibility of numbers. Instead of actually dividing into the numbers, try the rules in Table 2-2.

Table 2-2 Divisibility rules

A number is divisible by	*if*
2	it ends in 0,2,4,6, or 8
3	the sum of its digits is divisible by 3
4	the number formed by the last two digits is divisible by 4
5	it ends in 0 or 5
6	it is divisible by 2 and 3 (use the rules for both)
7	(no simple rule)
8	the number formed by the last three digits is divisible by 8
9	the sum of its digits is divisible by 9

Example 10: Answer the following.

(a) Is 126 divisible by 3?

(b) Is 1,648 divisible by 4?

(c) Is 186 divisible by 6?

(d) Is 2,488 divisible by 8?

(e) Is 2,853 divisible by 9?

(f) 4,620 is divisible by which of the following numbers?

2, 3, 4, 5, 6, 7, 8, 9.

Answers

(a) Sum of digits of 126 = 9. Since 9 is divisible by 3, 126 is divisible by 3.

(b) 48 is divisible by 4, so 1,648 is divisible by 4.

(c) 186 ends in 6, so it is divisible by 2. Sum of digits = 15.

15 is divisible by 3, so 186 is divisible by 3. 186 is divisible by 2 and 3; therefore, it is divisible by 6.

(d) 488 is divisible by 8, so 2,488 is divisible by 8.

(e) Sum of digits of 2,853 = 18. 18 is divisible by 9, so 2,853 is divisible by 9.

(f) 4,620 *is* divisible by

2—The number is even.

3—The sum of the digits is 12, which is divisible by 3.

4—The number formed by the last two digits, 20, is divisible by 4.

5—The number ends in 0.

6—The number is divisible by 2 and 3.

7—Divide 4,620 by 7, and you get 660.

4,620 is *not* divisible by

8—The number formed by the last three digits, 620, is not divisible by 8.

9—The sum of digits is 12, which is not divisible by 9.

Factors, Primes, Composites, and Factor Trees

You should become familiar with the definitions of certain types of numbers and how they can be found.

Factors

Numbers that are multiplied together to get a product are called **factors.**

Example 11: What are the factors of 18?

$$\text{factor} \times \text{factor} = 18$$

$$1 \times 18 = 18$$

$$2 \times 9 = 18$$

$$3 \times 6 = 18$$

So, the factors of 18 are 1, 2, 3, 6, 9, and 18. These numbers are also called the **divisors** of 18. *Factors* of a number are also called *divisors* of that same number.

Prime numbers

A **prime number** is a natural number, greater than 1, that can be divided by only itself and 1. Another definition: A prime number is a positive integer that has exactly two different factors: itself and 1.

Example 12: Is 19 a prime number?

Yes. The only factors of 19 are 1 and 19, so 19 is a prime number. That is, 19 is divisible by only 1 and 19, so it is prime.

Example 13: Is 27 a prime number?

No. 27 is divisible by other numbers (3 and 9), so it is not prime. The factors of 27 are 1, 3, 9, and 27, so it is not prime.

The only even prime number is 2; thereafter, any even number may be divided by 2. The numbers 0 and 1 are not prime numbers. The prime numbers less than 50 are 2, 3, 5, 7, 11, 13, 17, 19, 23, 29, 31, 37, 41, 43, and 47.

Composite numbers

A **composite number** is a natural number divisible by more than just 1 and itself. Another definition: A composite number is a positive integer that has more than two different factors. The numbers 4, 6, 8, 9, 10, 12, 14, 15, 16, 18, . . . are composite numbers because they are "composed" of other numbers. The numbers 0 and 1 are not composite numbers. (They are neither prime nor composite.)

Example 14: Is 25 a composite number?

Yes. 25 is divisible by 5, so it is composite. The factors of 25 are 1, 5, and 25.

Factor trees

Every composite number can be expressed as a product of prime factors. You can find **prime factors** by using a factor tree. A factor tree looks like this.

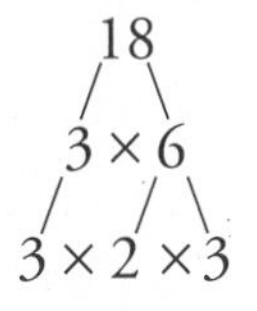

You can also make the tree as shown in the next tree.

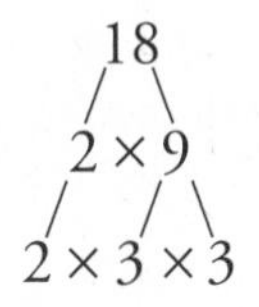

In either case, no matter how 18 is factored, the product of the primes is the same, even though the order may be different.

Example 15: Use a factor tree to express 60 as a product of prime factors.

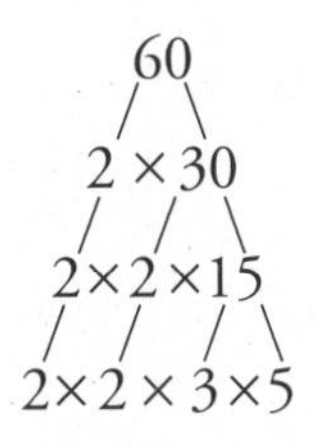

So the **prime factorization** of 60 is $2 \times 2 \times 3 \times 5$, which can be written as $2^2 \times 3 \times 5$. The actual *prime factors* of 60 are 2, 3, and 5.

Chapter Check-Out

1. Write 5,179 in expanded notation.
2. Round off 24,567 to the nearest thousand.
3. Estimate the sum of 2,455 + 5,644 rounded to the nearest thousand.
4. Estimate the product of 844 × 222 by rounding to the nearest hundred.
5. 7,686 is divisible by which of the following numbers?

   2, 3, 4, 5, 6, 7, 8, 9
6. Express 75 as a product of prime factors.

Answers: 1. $(5 \times 1{,}000) + (1 \times 100) + (7 \times 10) + (9 \times 1)$ or $(5 \times 10^3) + (1 \times 10^2) + (7 \times 10^1) + (9 \times 10^0)$ **2.** 25,000 **3.** 8,000 **4.** 160,000 **5.** 2, 3, 6, 7, 9 **6.** $3 \times 5 \times 5$ or 3×5^2

Chapter 3
DECIMALS

Chapter Check-In

- ❑ Place value
- ❑ Rounding decimals
- ❑ Decimal computation
- ❑ Estimating answers
- ❑ Repeating decimals

The system of numbers that you use is called the **decimal system** and is based on powers of ten (**base ten system**). Each place in the place value grid is ten times the value of the place to the right of it. Notice this version of the place value grid shown at the beginning of Chapter 2. Every number to the right of the decimal point is a **decimal fraction** (a fraction with a denominator of 10, 100, 1,000, and so on).

On the place value grid (see the following chart), notice that $\frac{1}{10}$ can be written as ten to a negative exponent, 10^{-1}. Similarly, $\frac{1}{100}$ can be written as ten to a negative exponent, 10^{-2}. (See Chapter 7 for more information about negative exponents.)

millions		1,000,000	10^6
hundred thousands		100,000	10^5
ten thousands		10,000	10^4
thousands		1,000	10^3
hundreds		100	10^2
tens		10	10^1
ones		1	10^0
tenths	1/10	0.1	10^{-1}
hundredths	1/100	0.01	10^{-2}
thousandths	1/1,000	0.001	10^{-3}
ten thousandths	1/10,000	0.0001	10^{-4}
hundred thousandths	1/100,000	0.00001	10^{-5}

Using the Place Value Grid

The place value grid can be used to assist you in understanding and working with the decimal system.

Expanded notation

Decimals can also be written in expanded notation, using the same techniques as when expanding whole numbers.

Example 1: Write 0.365 in expanded notation.

$$0.365 = 0.3 + 0.06 + 0.005$$

$$= (3 \times 0.1) + (6 \times 0.01) + (5 \times 0.001)$$

$$= (3 \times 10^{-1}) + (6 \times 10^{-2}) + (5 \times 10^{-3})$$

Example 2: Write 5.26 in expanded notation.

$$\begin{aligned}5.26 &= 5 + 0.2 + 0.06\\ &= (5 \times 1) + (2 \times 0.1) + (6 \times 0.01)\\ &= (5 \times 10^{0}) + (2 \times 10^{-1}) + (6 \times 10^{-2})\end{aligned}$$

Writing decimals

To read a decimal or write a decimal in words, you start at the left and end with the place value of the last number on the right. Where a whole number is included, use the word "and" to show the position of the decimal point.

Example 3: Read the number 0.75.

seventy-five hundredths

Example 4: Read the number 45.321.

forty-five and three hundred twenty-one thousandths

Example 5: Write two hundred and three tenths.

200.3

Comparing decimals

If you want to compare decimals, that is, find out whether one decimal is greater than another, simply make sure that each decimal goes out to the same number of places to the right.

Example 6: Which is greater, 0.37 or 0.365?

0.37 = 0.370, so you can align the two decimals.

0.370

0.365

It is easy to see that 0.37 is greater. You are really comparing three hundred seventy thousandths to three hundred sixty-five thousandths.

Example 7: Put the decimals 0.66, 0.6587, and 0.661 in order from largest to smallest.

First, change each number to ten-thousandths by adding zeros where appropriate. Then align the decimal points to make the comparison.

0.6600

0.6587

0.6610

The order should be 0.661, 0.66, and 0.6587.

You can also align the decimals first and then add the zeros as follows.

0.66 = 0.6600

0.6587 = 0.6587

0.661 = 0.6610

Remember: The number of digits to the right of the decimal point does not determine the size of the number (0.5 is greater than 0.33).

Rounding decimals

The method for rounding decimals is almost identical to the method used for rounding whole numbers. Follow these steps to round off a decimal:

1. Underline the place value to which you're rounding.
2. Look to the immediate right (one place) of your underlined place value.
3. Identify the number (the one to the right). If it is 5 or higher, round your underlined place value up 1 and drop all the numbers to the right of your underlined number. If the number (the one to the right) is 4 or less, leave your underlined place value as it is and drop all the numbers to the right of your underlined number. (Note that you do not have to replace dropped digits with zeroes.)

Example 8: Round off 0.478 to the nearest hundredth.

$0.4\underline{7}8$ is rounded up to 0.48.

Example 9: Round off 5.3743 to the nearest thousandth.

$5.37\underline{4}3$ is rounded down to 5.374

Decimal Computation

You should become efficient in using the four basic operations involving decimals—addition, subtraction, multiplication, and division.

Adding and subtracting decimals

To add or subtract decimals, just line up the decimal points and then add or subtract in the same manner you would add or subtract whole numbers.

Example 10: Add 23.6 + 1.75 + 300.002.

$$\begin{array}{r} 23.6 \\ 1.75 \\ +300.002 \\ \hline 325.352 \end{array}$$

Adding in zeros can make the problem easier to work.

$$\begin{array}{r} 23.600 \\ 1.750 \\ +300.002 \\ \hline 325.352 \end{array}$$

Example 11: Subtract 54.26 – 1.1.

$$\begin{array}{r} 54.26 \\ -\ \ 1.10 \\ \hline 53.16 \end{array}$$

Example 12: Subtract 78.9 – 37.43.

$$\begin{array}{r} 78.\overset{8}{\not{9}}\,{}^{1}0 \\ -\,37.4\ 3 \\ \hline 41.4\ 7 \end{array}$$

A whole number has an understood decimal point to its right.

Example 13: Subtract 17 – 8.43.

$$\begin{array}{r} 1\overset{6}{\not{7}}.\overset{9}{\not{0}}\,{}^{1}0 \\ -\ \ 8.\ 4\ 3 \\ \hline 8.\ 5\ 7 \end{array}$$

Multiplying decimals

To multiply decimals, just multiply as usual. Then count the total number of digits above the line that are to the right of all decimal points. Place your decimal point in your answer so there are the same number of digits to the right of it as there are above the line.

Example 14: Multiply 40.012×3.1.

$$\begin{array}{rl} 40.012 & \leftarrow 3\text{ digits} \\ \times \quad 3.1 & \leftarrow 1\text{ digit} \left\{ \text{total of 4 digits above the line that are to the right of the decimal point} \right. \\ \hline 40012 & \\ 120036 & \\ \hline 124.0372 & \leftarrow 4\text{ digits} \left\{ \text{decimal point placed so there are the same number of digits to the right of the decimal point} \right. \end{array}$$

Dividing decimals

Dividing decimals is the same as dividing other numbers, except that if the divisor (the number you're dividing by) has a decimal, move it to the right as many places as necessary until it is a whole number. Then move the decimal point in the dividend (the number being divided into) the same number of places. Sometimes, you may have to add zeros to the dividend (the number inside the division bracket). Note the decimal point in the quotient (answer) is placed above the one in the dividend.

Example 15: Divide $1.25\overline{)5.}$

$$1.25\overline{)5.} = 125\overline{)500.}\quad \text{quotient: } 4.$$

Example 16: Divide $0.002\overline{)26.}$

$$0.002\overline{)26.} = 2\overline{)26000.}\quad \text{quotient: } 13000.$$

Example 17: Divide $20\overline{)13.}$

$$
\begin{array}{r}
0.65 \\
20\overline{)13.} = 20\overline{)13.00} \\
\underline{120} \\
100 \\
\underline{100} \\
0
\end{array}
$$

Estimating Sums, Differences, Products, and Quotients

When working with decimals, it is easy to make a simple mistake and misplace the decimal point. Estimating an answer can be a valuable tool in helping you avoid this type of mistake.

Estimating sums

Use rounded numbers to estimate sums.

Example 18: Give an estimate for the sum of 19.61 and 5.07 by rounding to the nearest tenth.

Round each number to the nearest tenth.

$$
\begin{array}{c}
19.61 + 5.07 \\
\downarrow \quad \downarrow \\
19.6 + 5.1
\end{array}
$$

$$
\begin{array}{r}
19.6 \\
\underline{+\,5.1} \\
24.7
\end{array}
$$

$$\text{So } 19.61 + 5.07 \approx 24.7$$

Example 19: Estimate the sum of 19.61 + 5.07 by rounding to the nearest whole number.

Round each number to a whole number.

$$\begin{array}{ccccc} 19.61 & + & 5.07 & & \\ \downarrow & & \downarrow & & \\ 20 & + & 5 & = & 25 \end{array}$$

$$\text{So } 19.61 + 5.07 \approx 25$$

Estimating differences

Use rounded numbers to estimate differences.

Example 20: Give an estimate for the difference of 12.356 – 5.281 by rounding to the nearest whole number.

Round each number to the nearest whole number.

$$\begin{array}{ccc} 12.356 & - & 5.281 \\ \downarrow & & \downarrow \\ 12 & - & 5 \end{array}$$

Now subtract.

$$\begin{array}{r} 12 \\ -5 \\ \hline 7 \end{array}$$

So 12.356 – 5.281 ≈ 7.

Estimating products

Use rounded numbers to estimate products.

Example 21: Estimate the product of 4.7 × 5.9 by rounding to the nearest whole number.

Round each number to a whole number.

$$\begin{array}{c} 4.7 \times 5.9 \\ \downarrow \quad \downarrow \\ 5 \times 6 = 30 \end{array}$$

So $4.7 \times 5.9 \approx 30$.

Again, in decimals, as in whole numbers, if both multipliers end in .5, or are halfway numbers, rounding one number up and one number down will give you a better estimate of the product.

Example 22: Estimate the product of 7.5×8.5 by rounding to the nearest whole number.

$$\begin{array}{rc} & 7.5 \times 8.5 \\ & \downarrow \quad \downarrow \\ \text{Round one number up and one down} & 8 \times 8 = 64 \\ \text{So} & 7.5 \times 8.5 \approx 64 \end{array}$$

You can also round the first number down and the second number up and get this estimate.

$$\begin{array}{rc} & 7.5 \times 8.5 \\ & \downarrow \quad \downarrow \\ \text{Round one number up and one down} & 7 \times 9 = 63 \\ \text{So} & 7.5 \times 8.5 \approx 63 \end{array}$$

In either case, your approximation will be closer than it would be if you rounded both numbers up, which is the standard rule.

Estimating quotients

Use rounded numbers to estimate quotients.

Example 23: Estimate the quotient of $27.49 \div 3.12$ by rounding to the nearest whole number.

Round each number to the nearest whole number.

$$\begin{array}{rc} & 27.49 \div 3.12 \\ & \downarrow \quad \downarrow \\ & 27 \div 3 = 9 \\ \text{So} & 27.49 \div 3.12 \approx 9 \end{array}$$

Repeating Decimals

The most commonly used decimals are terminating decimals (decimals that stop, such as 0.5 or 0.74). A repeating decimal is a decimal that continues on indefinitely and repeats a number or block of numbers in a consistent manner, such as 0.666. . . or 0.232323. . .. A vinculum (a horizontal line over the number or numbers) is the standard notation used to show that a number or group of numbers is repeating. Using the vinculum, the repeating decimal looks like $0.\overline{6}$ or $0.\overline{23}$. Some books put the vinculum below the number, but this is less common.

Chapter Check-Out

1. Write 4.75 in expanded notation.
2. Write three hundred forty-seven and sixteen hundredths.
3. Which is greater, 0.44 or 0.436?
4. Round off 0.6778 to the nearest thousandth.
5. Add 51.3 + 1.2 + 16.003.
6. Subtract 23.6 – 2.77.
7. Multiply 22.3 × 1.1.
8. Divide 15.5 by 0.05.
9. Give an estimate for the difference of 23.55 – 8.41 by rounding to the nearest whole number.
10. Estimate the product of 6.5 × 3.5 by rounding to the nearest whole number.

Answers: 1. $(4 \times 1) + (7 \times 0.1) + (5 \times 0.01)$ or $(4 \times 10^0) + (7 \times 10^{-1}) + (5 \times 10^{-2})$ **2.** 347.16 **3.** 0.44 **4.** 0.678 **5.** 68.503 **6.** 20.83 **7.** 24.53 **8.** 310 **9.** 16 **10.** 21 or 24

Chapter 4
FRACTIONS

Chapter Check-In

- ❑ Proper and improper fractions
- ❑ Mixed numbers
- ❑ Renaming fractions
- ❑ Factors and multiples
- ❑ Operations with fractions and mixed numbers
- ❑ Simplifying fractions and complex fractions
- ❑ Changing fractions and decimals

A **fraction,** or fractional number, is used to represent a part of a whole. Fractions consist of two numbers: a **numerator** (which is above the line) and a **denominator** (which is below the line).

$$\frac{1}{2} \quad \begin{array}{l}\text{numerator}\\ \text{denominator}\end{array}$$

or

1/2 numerator/denominator

The denominator tells you the number of equal parts into which something is divided. The numerator tells you how many of these equal parts are being considered. Thus, if the fraction is $\frac{3}{5}$ of a pie, the denominator 5 tells you that the pie has been divided into 5 equal parts, of which 3 (numerator) are in the fraction. Sometimes, it helps to think of the *dividing line* (the middle of a fraction) as meaning "out of." In other words, $\frac{3}{5}$ also means 3 out of 5 equal pieces from the whole pie.

Proper and Improper Fractions

A fraction like $\frac{3}{5}$, where the numerator is smaller than the denominator, is less than one. This kind of fraction is called a **proper fraction.** But sometimes a fraction may be more than one or equal to one. This is when the numerator is larger than the denominator or equal to the denominator. Thus, $\frac{12}{7}$ is more than one and $\frac{6}{6}$ is equal to one. These kinds of fractions are called **improper fractions.**

- Examples of proper fractions: $\frac{4}{7}, \frac{2}{5}, \frac{1}{9}, \frac{10}{12}$
- Examples of improper fractions: $\frac{7}{4}, \frac{3}{2}, \frac{10}{3}, \frac{16}{15}, \frac{12}{12}$

Mixed Numbers

When a term contains both a whole number (3, 8, 25, and so on) and a fraction $\left(\frac{1}{2}, \frac{1}{4}, \frac{3}{4}, \text{ and so on}\right)$, the term is called a **mixed number.** For instance, $5\frac{1}{4}$ and $290\frac{3}{4}$ are both mixed numbers.

Changing improper fractions

To change an improper fraction to a mixed number, you divide the denominator into the numerator.

Example 1: Change $\frac{10}{3}$ to a mixed number.

$$\begin{array}{r} 3 \\ 3\overline{)10} \\ \underline{9} \\ 1 \end{array} \text{ remainder}$$

$$\frac{10}{3} = 3\frac{1}{3}$$

Changing mixed numbers

To change a mixed number to an improper fraction, you multiply the denominator times the whole number, add in the numerator, and put the total over the original denominator.

Example 2: Change $5\frac{3}{4}$ to an improper fraction.

$$4 \times 5 + 3 = 23$$

$$5\frac{3}{4} = \frac{23}{4}$$

Renaming Fractions

The following section reviews the processes of renaming and simplifying fractions. These processes are invaluable when you add, subtract, multiply, or divide fractions.

Equivalent fractions

Fractions that name the same number, such as 1/2, 2/4, 3/6, 4/8, and 5/10, are called equivalent fractions. A simple method to check if fractions are equivalent is to cross multiply and check the products.

Example 3: Is $\frac{2}{4}$ equivalent to $\frac{3}{6}$?

$$\frac{2}{4} \times \frac{3}{6}$$

$$12 = 12$$

The cross products are the same, so the fractions $\frac{2}{4}$ and $\frac{3}{6}$ are equivalent.

Example 4: Is $\frac{3}{4}$ equivalent to $\frac{2}{3}$?

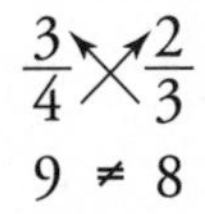

$$9 \neq 8$$

The cross products are not the same, so the fractions $\frac{3}{4}$ and $\frac{2}{3}$ are not equivalent.

Simplifying fractions

When given as a final answer, a fraction should be simplified to lowest terms. **Simplifying** fractions is done by dividing both the numerator and denominator by the largest integer that will divide evenly into both.

Example 5: Simplify $\frac{15}{25}$ to lowest terms.

To simplify $\frac{15}{25}$ to lowest terms, divide the numerator and denominator by 5.

$$\frac{15}{25} = \frac{15 \div 5}{25 \div 5} = \frac{3}{5}$$

Since $\frac{3}{5}$ cannot be simplified any further, that is, the numerator and denominator cannot both be evenly divided again, $\frac{3}{5}$ is simplified to lowest terms.

Example 6: Simplify $\frac{8}{40}$ to lowest terms.

$$\frac{8}{40} = \frac{8 \div 8}{40 \div 8} = \frac{1}{5}$$

Enlarging denominators

The **denominator** of a fraction may be enlarged and the fraction keeps its original value by multiplying both the numerator and denominator by the same number.

Example 7: Change $\frac{3}{4}$ to eighths.

To change $\frac{3}{4}$ to eighths, simply multiply the numerator and denominator by 2.

$$\frac{3}{4} = \frac{3 \times 2}{4 \times 2} = \frac{6}{8}$$

Example 8: Express $\frac{1}{2}$ as tenths.

$$\frac{1}{2} = \frac{1 \times 5}{2 \times 5} = \frac{5}{10}$$

Factors

As mentioned earlier, **factors** of a number are whole numbers that when multiplied together yield the number.

Example 9: What are the factors of 10?

$$10 = 2 \times 5$$

$$\text{and } 10 = 1 \times 10$$

So the factors of 10 are 1, 2, 5, and 10.

Example 10: What are the factors of 24?

$$\begin{aligned} 24 &= 1 \times 24 \\ &= 2 \times 12 \\ &= 3 \times 8 \\ &= 4 \times 6 \end{aligned}$$

Therefore, the factors of 24 are 1, 2, 3, 4, 6, 8, 12, and 24.

Common factors

Common factors are those factors that are the same for two or more numbers.

Example 11: What are the common factors of 6 and 8?

Factors of 6:	(1) (2) 3 6
Factors of 8:	(1) (2) 4 8

1 and 2 are common factors of 6 and 8.

Note: Some numbers may have many common factors.

Example 12: What are the common factors of 24 and 36?

Factors of 24:	(1) (2) (3) (4) (6) 8 (12) 24
Factors of 36:	(1) (2) (3) (4) (6) 9 (12) 18 36

Thus, the common factors of 24 and 36 are 1, 2, 3, 4, 6, and 12.

Greatest common factor

The **greatest common factor (GCF)** is the largest factor common to two or more numbers.

Example 13: What is the greatest common factor of 12 and 30?

Factors of 12:	(1) (2) (3) 4 (6) 12
Factors of 30:	(1) (2) (3) 5 (6) 10 15 30

Notice that, although 1, 2, 3, and 6 are all common factors of 12 and 30, only 6 is the greatest common factor.

Multiples

Multiples of a number are found by multiplying that number by 1, by 2, by 3, by 4, by 5, etc.

Example 14: List the first seven multiples of 9.

9, 18, 27, 36, 45, 54, 63

Common multiples

Common multiples are multiples that are the same for two or more numbers.

Example 15: What are the common multiples of 2 and 3?

Multiples of 2:	2	4	6	8	10	12	14	16	18	etc.
Multiples of 3:		3	6	9		12	15		18	etc.

The common multiples of 2 and 3 are 6, 12, 18, …

Notice that common multiples may go on indefinitely.

Least common multiple

The **least common multiple (LCM)** is the smallest multiple that is common to two or more numbers.

Example 16: What is the least common multiple of 2 and 3?

Multiples of 2:	2	4	6	8	10	12	etc.
Multiples of 3:		3	6	9		12	etc.

The smallest multiple common to both 2 and 3 is 6.

Example 17: What is the least common multiple of 2, 3, and 4?

Multiples of 2:	2	4	6	8	10		12	etc.
Multiples of 3:		3	6		9		12	etc.
Multiples of 4:		4		8			12	etc.

The least common multiple of 2, 3, and 4 is 12.

Adding and Subtracting Fractions

As you review adding and subtracting fractions, notice the steps that are the same for both operations.

Adding fractions

To add fractions, you must have a common denominator. Fractions that have common denominators are called like fractions. Fractions that have different denominators are called unlike fractions. To add *like* fractions, simply add the numerators and keep the same (or like) denominator.

Example 18: Add $\frac{1}{5}+\frac{3}{5}$

$$\begin{array}{r} \frac{1}{5} \\ +\frac{3}{5} \\ \hline \frac{4}{5} \end{array}$$

To add *unlike* fractions, first change all denominators to their lowest common denominator (LCD), also called the lowest common multiple of the denominator, the lowest number that can be divided evenly by all denominators in the problem. The numerators may need to be changed to make sure that the fractions are still equivalent to the originals. When you have all the denominators the same, you may add the numerators and keep the same denominator.

Example 19: Add

(a) $\frac{3}{8}+\frac{1}{2}$

Change the $\frac{1}{2}$ to $\frac{4}{8}$ because the 8 is the lowest common denominator; then add the numerators 3 and 4 to get $\frac{7}{8}$.

$$\begin{array}{r} \frac{3}{8}=\frac{3}{8} \\ +\frac{1}{2}=\frac{4}{8} \\ \hline \frac{7}{8} \end{array} \leftarrow \left(\text{change } \frac{1}{2} \text{ to } \frac{4}{8}\right)$$

(b) $\frac{1}{4}+\frac{1}{3}$

Change both fractions to get the lowest common denominator of 12, and then add the numerators to get $\frac{7}{12}$.

$$\begin{array}{r} \frac{1}{4}=\frac{3}{12} \\ +\frac{1}{3}=\frac{4}{12} \\ \hline \frac{7}{12} \end{array} \leftarrow \text{(change both fractions to LCD of 12)}$$

Note: Fractions may be added across as well.

Example 20: Add $\frac{1}{2}+\frac{1}{3}$.

$$\frac{1}{2}+\frac{1}{3}=\frac{3}{6}+\frac{2}{6}=\frac{5}{6}$$

Subtracting fractions

To subtract fractions, the same rule as in adding fractions applies (find the LCD), except that you subtract the numerators.

Example 21: Subtract

(a) $\frac{7}{8}-\frac{1}{4}$

$$\begin{array}{r} \frac{7}{8}=\frac{7}{8} \\ -\frac{1}{4}=\frac{2}{8} \\ \hline \frac{5}{8} \end{array}$$

(b) $\frac{3}{4}-\frac{1}{3}$

$$\begin{array}{r} \frac{3}{4} = \frac{9}{12} \\ -\frac{1}{3} = \frac{4}{12} \\ \hline \frac{5}{12} \end{array}$$

Again, a subtraction problem may be done across as well as down.

Example 22: Subtract $\frac{5}{8} - \frac{3}{8}$

$$\frac{5}{8} - \frac{3}{8} = \frac{2}{8} = \frac{1}{4}$$

Adding and Subtracting Mixed Numbers

You will notice that the steps in adding and subtracting mixed numbers are similar. But also note the differences and the reminders given in this section.

Adding mixed numbers

To add mixed numbers, the same rule as in adding fractions applies (find the LCD), but make sure that you always add the *whole numbers* to get your final answer.

Example 23: Add $2\frac{1}{2} + 3\frac{1}{4}$.

$$\begin{array}{r} 2\frac{1}{2} = 2\frac{2}{4} \leftarrow \left(\frac{1}{2} \text{ is changed to } \frac{2}{4}\right) \\ +\,3\frac{1}{4} = 3\frac{1}{4} \\ \hline 5\frac{3}{4} \\ \uparrow \end{array}$$

(remember to add the whole numbers)

Sometimes, you may end up with a mixed number that includes an improper fraction. In that case, you must change the improper fraction to a mixed number and combine it with the sum of the integers.

Example 24: Add $2\frac{1}{2}+5\frac{3}{4}$.

$$\begin{array}{r} 2\frac{1}{2}=2\frac{2}{4} \\ +5\frac{3}{4}=5\frac{3}{4} \\ \hline 7\frac{5}{4} \end{array}$$

And $\frac{5}{4}=1\frac{1}{4}$

So $7\frac{5}{4}=7+1\frac{1}{4}=8\frac{1}{4}$

Subtracting mixed numbers

When you subtract mixed numbers, you sometimes may have to rename the whole number, just as you sometimes borrow from the next column when subtracting whole numbers. ***Note:*** When you borrow 1 from the whole number, the 1 must be changed to a fraction.

Example 25: Subtract

(a) 651 – 129 **(b)** $4\frac{1}{6}-2\frac{5}{6}$ **(c)** $5\frac{1}{5}-3\frac{1}{2}$

(a)

$$\begin{array}{r} 6\overset{4}{\not{5}}\overset{11}{1} \\ -129 \\ \hline 522 \end{array}$$

(You borrowed 1 from the 10's column.)

(b)

$$\begin{array}{rcr} 4\frac{1}{6} & \rightarrow & 3\frac{7}{6} \\ -2\frac{5}{6} & \rightarrow & -2\frac{5}{6} \\ \hline & & 1\frac{2}{6}=1\frac{1}{3} \end{array}$$

(You borrowed 1 in the form of $\frac{6}{6}$ from the 1's column.)

$$
\begin{array}{rl}
& 5\frac{1}{5} = 5\frac{2}{10} = 4\frac{12}{10} \\
\text{(c)} & \underline{-3\frac{1}{2} = 3\frac{5}{10} = 3\frac{5}{10}} \\
& \qquad\qquad\qquad 1\frac{7}{10}
\end{array}
$$

(You borrowed 1 in the form of $\frac{10}{10}$ from the 1's column.)

Notice that you should borrow only after you have identified a common denominator.

To subtract a mixed number from a whole number, you have to borrow from the whole number.

Example 26: Subtract $6 - 3\frac{1}{5}$.

$$
\begin{array}{r}
6 = 5\frac{5}{5} \leftarrow \text{(borrow 1 in the form of } \frac{5}{5} \text{ from the 6)} \\
\underline{-3\frac{1}{5} = 3\frac{1}{5}} \\
2\frac{4}{5} \\
\uparrow
\end{array}
$$

(remember to subtract the remaining whole numbers)

Multiplying Fractions and Mixed Numbers

The steps involved in multiplying fractions and mixed numbers are similar, but an important first step is required when multiplying mixed numbers.

Multiplying fractions

To multiply fractions, simply multiply the numerators; then multiply the denominators. Simplify to lowest terms if possible.

Example 27: Multiply $\frac{2}{3} \times \frac{5}{12}$.

$$\frac{2}{3} \times \frac{5}{12} = \frac{10}{36}$$

$$\text{Simplify} \quad \frac{10}{36} = \frac{5}{18}$$

This answer had to be simplified because it wasn't in lowest terms.

When multiplying fractions, you can *cancel* (do early simplifying) first, which eliminates the need to simplify your answer later. To cancel, find a number that divides evenly into one numerator and one denominator. In Example 27, 2 will divide evenly into 2 in the numerator (it goes in one time) and 12 in the denominator (it goes in six times). Thus,

$$\frac{\overset{1}{\cancel{2}}}{3} \times \frac{5}{\underset{6}{\cancel{12}}} = \frac{5}{18}$$

Whole numbers can also be written as fractions $\left(3 = \frac{3}{1} \text{ or } 4 = \frac{4}{1}\right)$, so the problem $3 \times \frac{3}{8}$ is worked by changing 3 to $\frac{3}{1}$.

Example 28: Multiply $3 \times \frac{3}{8}$.

$$\begin{aligned} 3 \times \frac{3}{8} &= \frac{3}{1} \times \frac{3}{8} \\ &= \frac{9}{8} \\ &= 1\frac{1}{8} \end{aligned}$$

Example 29: Multiply $\frac{1}{4} \times \frac{2}{7}$.

$$\frac{1}{\not{4}_{2}} \times \frac{\not{2}^{1}}{7} = \frac{1}{14}$$

Remember: You may cancel only when *multiplying* fractions.

Multiplying mixed numbers

To multiply mixed numbers, first change any mixed number to an improper fraction. Then multiply the numerators together and the denominators together, as shown in Example 27.

Example 30: Multiply $3\frac{1}{3} \times 2\frac{1}{4}$.

$$3\frac{1}{3} \times 2\frac{1}{4} = \frac{10}{3} \times \frac{9}{4} = \frac{90}{12} = 7\frac{6}{12} = 7\frac{1}{2}$$

Or $$\frac{\not{10}^{5}}{\not{3}_{1}} \times \frac{\not{9}^{3}}{\not{4}_{2}} = \frac{15}{2} = 7\frac{1}{2}$$

Example 31: Multiply $3\frac{1}{5} \times 6\frac{1}{2}$.

$$3\frac{1}{5} \times 6\frac{1}{2} = \frac{16}{5} \times \frac{13}{2} = \frac{\not{16}^{8}}{5} \times \frac{13}{\not{2}_{1}} = \frac{104}{5} = 20\frac{4}{5}$$

Dividing Fractions and Mixed Numbers

The steps involved in dividing fractions and mixed numbers are similar, but an important step is required when dividing mixed numbers.

Dividing fractions

To divide fractions, *invert* (turn upside down) the second fraction (the one "divided by") and multiply. Then simplify, if possible. Division of fractions can also be performed by multiplying the first fraction by the reciprocal of the second fraction.

Example 32: Divide

(a) $\frac{1}{6} \div \frac{1}{5}$ (b) $\frac{1}{6} \div \frac{1}{3}$

(a) $\frac{1}{6} \div \frac{1}{5} = \frac{1}{6} \times \frac{5}{1} = \frac{5}{6}$ (b) $\frac{1}{6} \div \frac{1}{3} = \frac{1}{\cancel{6}_2} \times \frac{\cancel{3}^1}{1} = \frac{1}{2}$

Example 33: Divide $\frac{1}{3} \div 6$.

$6 = \frac{6}{1}$, so the problem can be written $\frac{1}{3} \div \frac{6}{1}$. Then invert the second fraction and multiply.

$$\frac{1}{3} \div \frac{6}{1} = \frac{1}{3} \times \frac{1}{6} = \frac{1}{18}$$

Example 34: Divide $\frac{3}{7} \div \frac{3}{14}$.

$$\frac{3}{7} \div \frac{3}{14} = \frac{\cancel{3}^1}{\cancel{7}_1} \times \frac{\cancel{14}^2}{\cancel{3}_1} = \frac{2}{1} = 2$$

Dividing complex fractions

Sometimes, a division of fractions problem may appear in the following form. These are called complex fractions.

$$\frac{\frac{3}{4}}{\frac{7}{8}}$$

The line separating the two fractions means "divided by." Therefore, this problem may be rewritten as

$$\frac{3}{4} \div \frac{7}{8}$$

Now follow the same procedure as shown in Example 33.

$$\frac{3}{4} \div \frac{7}{8} = \frac{3}{\not{4}_1} \times \frac{\not{8}^2}{7} = \frac{6}{7}$$

Example 35: Divide = $\dfrac{\frac{7}{8}}{\frac{1}{2}}$

$$\frac{\frac{7}{8}}{\frac{1}{2}} = \frac{7}{8} \div \frac{1}{2} = \frac{7}{\not{8}_4} \times \frac{\not{2}^1}{1} = \frac{7}{4} = 1\frac{3}{4}$$

Dividing mixed numbers

To divide mixed numbers, first change them to improper fractions (Example 2). Then follow the rule for dividing fractions (Example 34).

Example 36: Divide

(a) $3\frac{3}{5} \div 2\frac{2}{3}$ **(b)** $2\frac{1}{5} \div 3\frac{1}{10}$

(a) $3\frac{3}{5} \div 2\frac{2}{3} = \frac{18}{5} \div \frac{8}{3} = \frac{\not{18}^9}{5} \times \frac{3}{\not{8}_4} = \frac{27}{20} = 1\frac{7}{20}$

(b) $2\frac{1}{5} \div 3\frac{1}{10} = \frac{11}{5} \div \frac{31}{10} = \frac{11}{\not{5}_1} \times \frac{\not{10}^2}{31} = \frac{22}{31}$

Notice that after you invert and have a multiplication of fractions problem, you may then cancel tops with bottoms when appropriate.

Simplifying Fractions and Complex Fractions

If either numerator or denominator consists of several numbers, these numbers must be combined into one number. Then simplify if possible.

Example 37: Simplify.

(a) $\dfrac{28+14}{26+17}$ (b) $\dfrac{\frac{1}{4}+\frac{1}{2}}{\frac{1}{3}+\frac{1}{4}}$ (c) $\dfrac{2+\frac{1}{2}}{3+\frac{1}{4}}$

(d) $\dfrac{3-\frac{3}{4}}{4+\frac{1}{2}}$ (e) $\dfrac{1}{1+\dfrac{1}{1+\frac{1}{4}}}$

(a) $\dfrac{28+14}{26+17}=\dfrac{42}{43}$

(b)
$$\begin{aligned}
\frac{\frac{1}{4}+\frac{1}{2}}{\frac{1}{3}+\frac{1}{4}} &= \frac{\frac{1}{4}+\frac{2}{4}}{\frac{4}{12}+\frac{3}{12}} \\
&= \frac{\frac{3}{4}}{\frac{7}{12}} \\
&= \frac{3}{4}\div\frac{7}{12} \\
&= \frac{3}{\cancel{4}_1}\times\frac{\cancel{12}^3}{7} \\
&= \frac{9}{7} \\
&= 1\frac{2}{7}
\end{aligned}$$

(c) $$\frac{2+\frac{1}{2}}{3+\frac{1}{4}} = \frac{2\frac{1}{2}}{3\frac{1}{4}}$$

$$= \frac{\frac{5}{2}}{\frac{13}{4}}$$

$$= \frac{5}{2} \div \frac{13}{4}$$

$$= \frac{5}{\cancel{2}_1} \times \frac{\cancel{4}^2}{13}$$

$$= \frac{10}{13}$$

(d) $$\frac{3-\frac{3}{4}}{4+\frac{1}{2}} = \frac{2\frac{1}{4}}{4\frac{1}{2}}$$

$$= \frac{\frac{9}{4}}{\frac{9}{2}}$$

$$= \frac{9}{4} \div \frac{9}{2}$$

$$= \frac{\cancel{9}^1}{\cancel{4}_2} \times \frac{\cancel{2}^1}{\cancel{9}_1}$$

$$= \frac{1}{2}$$

(e) $$\frac{1}{1+\frac{1}{1+\frac{1}{4}}} = \frac{1}{1+\frac{1}{1\frac{1}{4}}}$$

$$= \frac{1}{1+\frac{1}{\frac{5}{4}}}$$

$$= \frac{1}{1+\left(1 \div \frac{5}{4}\right)}$$

$$= \frac{1}{1+\left(\frac{1}{1} \times \frac{4}{5}\right)}$$

$$= \frac{1}{1+\frac{4}{5}}$$

$$= \frac{1}{1\frac{4}{5}}$$

$$= \frac{1}{\frac{9}{5}}$$

$$= \frac{1}{1} \div \frac{9}{5}$$

$$= \frac{1}{1} \times \frac{5}{9}$$

$$= \frac{5}{9}$$

Changing Fractions to Decimals

Fractions may also be written in **decimal** form **(decimal fractions)** as either terminating (for example, 0.3) or infinite repeating (for example, 0.66. . .) decimals. To change a fraction to a decimal, simply do what the operation says. In other words, $\frac{13}{20}$ means 13 divided by 20. Insert decimal points and zeros accordingly.

Example 38: Change to decimals.

(a) $\frac{5}{8}$ (b) $\frac{2}{5}$ (c) $\frac{4}{9}$

(a) $\frac{5}{8} = 8\overline{)5.000}$ = 0.625

$$\begin{array}{r} 0.625 \\ 8\overline{)5.000} \\ \underline{48} \\ 20 \\ \underline{16} \\ 40 \\ \underline{40} \\ 0 \end{array}$$

So $\frac{5}{8} = 0.625$

(b) $\frac{2}{5} = 5\overline{)2.0}$

$$\begin{array}{r} 0.4 \\ 5\overline{)2.0} \\ \underline{20} \\ 0 \end{array}$$

So $\frac{2}{5} = 0.4$

(c) $\frac{4}{9} = 9\overline{)4.000}$

$$\begin{array}{r} 0.444... \\ 9\overline{)4.000} \\ \underline{36} \\ 40 \\ \underline{36} \\ 40 \end{array}$$

So $\frac{4}{9} = 0.444...$ or $0.\overline{4}$

Changing Terminating Decimals to Fractions

To change terminating decimals to fractions, simply remember that all numbers to the right of the decimal point are fractions with denominators of only 10, 100, 1,000, 10,000, and so on. Next, use the technique of *read it*, *write it*, and *reduce it* (that is, simplify).

Example 39: Change the following to fractions in lowest terms.

(a) 0.8 **(b)** 0.09

(a) *Read it:* 0.8 (eight tenths)

Write it: $\frac{8}{10}$

Simplify it: $\frac{4}{5}$

(b) *Read it:* 0.09 (nine hundredths)

Write it: $\frac{9}{100}$

Simplify it: $\frac{9}{100}$ can't be simplified.

Changing Infinite Repeating Decimals to Fractions

Remember: Infinite repeating decimals are usually represented by putting a line over (sometimes under) the shortest block of repeating decimals. *Every infinite repeating decimal can be expressed as a fraction.*

Example 40: Find the fraction represented by the repeating decimal $0.\overline{7}$.

Let n stand for $0.\overline{7}$ or 0.77777...

So $10n$ stands for $7.\overline{7}$ or 7.77777...

$10n$ and n have the same fractional part, so their difference is an integer.

$$\begin{array}{r} 10n = 7.\overline{7} \\ -\ n = 0.\overline{7} \\ \hline 9n = 7 \end{array}$$

You can solve this problem as follows.

$$9n = 7$$

$$n = \frac{7}{9}$$

So $0.\overline{7} = \frac{7}{9}$

Example 41: Find the fraction represented by the repeating decimal $0.\overline{36}$.

Let n stand for $0.\overline{36}$ or 0.363636...

So $10n$ stands for $3.6\overline{36}$ or 3.63636...

and $100n$ stands for $36.\overline{36}$ or 36.3636...

$100n$ and n have the same fractional part, so their difference is an integer. (The repeating parts are the same, so they subtract out.)

$$\begin{array}{r} 100n = 36.\overline{36} \\ -\quad n = 0.\overline{36} \\ \hline 99n = 36 \end{array}$$

You can solve this equation as follows:

$$99n = 36$$

$$n = \frac{36}{99}$$

Now simplify $\frac{36}{99}$ to $\frac{4}{11}$.

So $\quad 0.\overline{36} = \frac{4}{11}$

Example 42: Find the fraction represented by the repeating decimal $0.5\overline{4}$.

Let n stand for $0.5\overline{4}$ or 0.544444...

So $10n$ stands for $5.\overline{4}$ or 5.444444...

and $100n$ stands for $54.\overline{4}$ or 54.4444...

Since $100n$ and $10n$ have the same fractional part, their difference is an integer. (Again, notice how the repeated parts must align to subtract out.)

$$\begin{array}{r} 100n = 54.\overline{4} \\ -10n = 5.\overline{4} \\ \hline 90n = 49 \end{array}$$

You can solve this equation as follows.

$$90n = 49$$

$$n = \frac{49}{90}$$

$$\text{So } 0.5\overline{4} = \frac{49}{90}$$

Chapter Check-Out

1. Change $\frac{5}{3}$ into a mixed number.
2. Change $6\frac{1}{3}$ into an improper fraction.
3. Is $\frac{2}{3}$ equivalent to $\frac{5}{9}$?
4. What is the least common multiple of 4 and 5?
5. Add $\frac{1}{5}+\frac{2}{7}$.
6. Add $3\frac{1}{6}+2\frac{3}{4}$.
7. Subtract $5\frac{1}{3}-2\frac{3}{7}$.
8. Multiply $\frac{3}{5}\times\frac{2}{7}$.
9. Multiply $5\frac{1}{2}\times2\frac{1}{3}$.
10. Divide $4\frac{1}{2}$ by $2\frac{1}{4}$.
11. Change $\frac{3}{8}$ to a decimal.
12. Change 0.454545… to a fraction.

Answers: 1. $1\frac{2}{3}$ **2.** $\frac{19}{3}$ **3.** No **4.** 20 **5.** $\frac{17}{35}$ **6.** $5\frac{11}{12}$ **7.** $2\frac{19}{21}$ **8.** $\frac{6}{35}$ **9.** $\frac{77}{6}$ or $12\frac{5}{6}$ **10.** 2 **11.** 0.375 **12.** $\frac{5}{11}$

Chapter 5
PERCENTS

Chapter Check-In

- ❑ Changing percents, decimals, and fractions
- ❑ Important equivalents
- ❑ Applications of percents
- ❑ Solving percent problems

A fraction whose denominator is 100 is called a **percent.** The word *percent* means hundredths (per hundred).

So $43\% = \frac{43}{100}$.

Changing Percents, Decimals, and Fractions

Using the step-by-step methods of changing decimals, percents, and fractions makes the conversions straightforward and simple to complete.

Changing decimals to percents

To change decimals to percents,

1. Move the decimal point two places to the right.
2. Insert a percent sign.

Example 1: Change to percents.

(a) 0.55 **(b)** 0.02 **(c)** 3.21 **(d)** 0.004

(a) 0.55 = 55%
(b) 0.02 = 2%
(c) 3.21 = 321%
(d) 0.004 = 0.4%

Changing percents to decimals

To change percents to decimals,

1. Eliminate the percent sign.
2. Move the decimal point two places to the left. (Sometimes, adding zeros is necessary.)

Example 2: Change to decimals.

(a) 24% **(b)** 7% **(c)** 250%

(a) 24% = 0.24
(b) 7% = 0.07
(c) 250% = 2.50 = 2.5

Changing fractions to percents

To change fractions to percents,

1. Change to a decimal.
2. Change the decimal to a percent.

Example 3: Change to percents.

(a) $\frac{1}{2}$ **(b)** $\frac{2}{5}$ **(c)** $\frac{5}{2}$ **(d)** $\frac{1}{20}$

(a) $\frac{1}{2} = 0.5 = 50\%$

(b) $\frac{2}{5} = 0.4 = 40\%$

(c) $\frac{5}{2} = 2.5 = 250\%$

(d) $\frac{1}{20} = 0.05 = 5\%$

Changing percents to fractions

There are two simple methods for changing percents to fractions.

Method 1.

1. Drop the percent sign.
2. Write over one hundred.
3. Simplify if possible.

Example 4(a): (Applying method 1) Change to fractions.

(a) 13% **(b)** 70% **(c)** 45% **(d)** 130%

(a) $13\% = \frac{13}{100}$

(b) $70\% = \frac{70}{100} = \frac{7}{10}$

(c) $45\% = \frac{45}{100} = \frac{9}{20}$

(d) $130\% = \frac{130}{100} = \frac{13}{10} = 1\frac{3}{10}$

Method 2.

1. Drop the percent sign.
2. Multiply by $\frac{1}{100}$.
3. Simplify if possible.

Example 4(b): (Applying method 2) Change to fractions.

$$\text{(a) } 66\frac{2}{3}\% \rightarrow 66\frac{2}{3}\times\frac{1}{100}=\frac{\overset{2}{\cancel{200}}}{3}\times\frac{1}{\underset{1}{\cancel{100}}}=\frac{2}{3}$$

$$\text{(b) } 112\frac{1}{2}\% \rightarrow 112\frac{1}{2}\times\frac{1}{100}=\frac{\overset{9}{\cancel{225}}}{2}\times\frac{1}{\underset{4}{\cancel{100}}}=\frac{9}{8}\text{ or } 1\frac{1}{8}$$

Important Equivalents

Important equivalents can save you time. Memorizing the equivalents that follow can eliminate computations.

$\frac{1}{100}=0.01=1\%$

$\frac{1}{10}=0.1=10\%$

$\frac{2}{10}=\frac{1}{5}=0.2=20\%$

$\frac{3}{10}=0.3=30\%$

$\frac{4}{10}=\frac{2}{5}=0.4=40\%$

$\frac{5}{10}=\frac{1}{2}=0.5=50\%$

$\frac{6}{10}=\frac{3}{5}=0.6=60\%$

$\frac{7}{10}=0.7=70\%$

$\frac{8}{10}=\frac{4}{5}=0.8=80\%$

$\frac{9}{10}=0.9=90\%$

$\frac{1}{8}=0.125=12.5\%=12\frac{1}{2}\%$

$\frac{2}{8}=\frac{1}{4}=0.25=25\%$

$\frac{3}{8}=0.375=37.5\%=37\frac{1}{2}\%$

$\frac{4}{8}=\frac{1}{2}=0.5=50\%$

$\frac{5}{8}=0.625=62.5\%=62\frac{1}{2}\%$

$\frac{6}{8}=\frac{3}{4}=0.75=75\%$

$\frac{7}{8}=0.875=87.5\%=87\frac{1}{2}\%$

$\frac{1}{6} = 0.16\frac{2}{3} = 16\frac{2}{3}\%$

$\frac{2}{6} = \frac{1}{3} = 0.33\frac{1}{3} = 33\frac{1}{3}\%$

$\frac{3}{6} = \frac{1}{2} = 0.5 = 50\%$

$\frac{4}{6} = \frac{2}{3} = 0.66\frac{2}{3} = 66\frac{2}{3}\%$

$\frac{5}{6} = 0.83\frac{1}{3} = 83\frac{1}{3}\%$

$1 = 1.0 = 100\%$

$2 = 2.0 = 200\%$

$3\frac{1}{2} = 3.5 = 350\%$

Applications of Percents

Percents can be used in many types of problems and situations. The following applications are the most common basic types.

Finding percent of a number

To determine percent of a number, change the percent to a fraction or decimal (whichever is easier for you) and multiply. ***Remember:*** The word *of* means multiply.

Example 5: Find the percents of the following numbers.

(a) What is 20% of 80? **(b)** What is 15% of 50?

(c) What is $\frac{1}{2}$% of 18? **(d)** What is 70% of 20?

(a) Using fractions,

$$20\% \text{ of } 80 = \frac{\overset{1}{\cancel{20}}}{\underset{\underset{1}{\cancel{5}}}{\cancel{100}}} \times \frac{\overset{16}{\cancel{80}}}{1} = 16$$

Using decimals,

20% of 80 = $0.20 \times 80 = 16.00 = 16$

(b) Using fractions,

$$15\% \text{ of } 50 = \frac{\overset{3}{\cancel{15}}}{\underset{\underset{2}{\cancel{20}}}{\cancel{100}}} \times \frac{\overset{5}{\cancel{50}}}{1} = \frac{15}{2} = 7\frac{1}{2} \text{ or } 7.5$$

Using decimals,

$15\% \text{ of } 50 = 0.15 \times 50 = 7.5$

(c) Using fractions,

$$\frac{1}{2}\% \text{ of } 18 = \frac{1}{\underset{1}{\cancel{2}}} \times \frac{1}{100} \times \frac{\overset{9}{\cancel{18}}}{1} = \frac{9}{100} \text{ or } 0.09$$

Using decimals,

$$\frac{1}{2}\% \text{ of } 18 = 0.005 \times 18 = 0.09$$

(d) Using fractions,

$$70\% \text{ of } 20 = \frac{\overset{7}{\cancel{70}}}{\underset{\underset{1}{\cancel{10}}}{\cancel{100}}} \times \frac{\overset{2}{\cancel{20}}}{1} = \frac{14}{1} = 14$$

Using decimals,

$70\% \text{ of } 20 = 0.70 \times 20 = 14$

Finding what percent one number is of another

One method to find what percent one number is of another is the division method. To use this method, simply take the number after the *of* and divide it into the number next to the *is.* Then change the answer to a percent.

Example 6: Find the percentages.

(a) 20 is what percent of 50?

(b) 27 is what percent of 90?

(a) $\frac{20}{50} = \frac{2}{5} = 0.4 = 40\%$

(b) $\frac{27}{90} = \frac{3}{10} = 0.3 = 30\%$

Another method to find what percent one number is of another is the equation method. Simply turn the question word-for-word into an equation. (To review solving simple equations, see Chapter 13). For *what,* substitute the letter *x;* for *is*, substitute an *equal sign* (=); for *of,* substitute a *multiplication sign* (×). Change percents to decimals or fractions, whichever you find easier. Then solve the equation.

Example 7: Change each of the following into an equation and solve.

(a) 10 is what percent of 50?

(b) 15 is what percent of 60?

(a) 10 is what percent of 50?

$$10 = x(50)$$

$$\frac{10}{50} = \frac{x(50)}{50}$$

$$\frac{10}{50} = \frac{x(\overset{1}{\cancel{50}})}{\underset{1}{\cancel{50}}}$$

$$\frac{10}{50} = x$$

$$\frac{1}{5} = x$$

$$20\% = x$$

Therefore, 10 is 20% of 50.

(b) 15 is what percent of 60?

$$15 = x(60)$$
$$\frac{15}{60} = \frac{x(60)}{60}$$
$$\frac{15}{60} = \frac{x(\cancel{60}^{1})}{\cancel{60}_{1}}$$
$$\frac{15}{60} = x$$
$$\frac{1}{4} = x$$
$$25\% = x$$

Therefore, 15 is 25% of 60.

Finding a number when a percent of it is known

You can also use the division method to find a number when a percent of it is known. To use this method, simply take the number of percent, change it into a decimal or fraction, and divide that into the other number.

You could also use the equation method as discussed previously. Simply turn the question word-for-word into an equation. For *what,* substitute the letter *x;* for *is,* substitute an *equal sign* (=); for *of,* substitute a *multiplication sign* (×). Change percents to decimals or fractions, whichever you find easier. Then solve the equation.

Example 8: Find the number.

(a) 15 is 50% of what number?

(b) 20 is 40% of what number?

(a) $15 = 50\%x \text{ or } \frac{15}{0.50} = 30 \text{ or } \frac{15}{\frac{1}{2}} = \frac{15}{1} \times \frac{2}{1} = 30$

(b) $\frac{20}{0.40} = 50 \text{ or } \frac{20}{\frac{2}{5}} = \frac{\cancel{20}^{10}}{1} \times \frac{5}{\cancel{2}_{1}} = 50$

Example 9: Find the number.

(a) 30 is 20% of what number?

(b) 40 is 80% of what number?

(a) 30 is 20% of what number?

$$30 = 0.20(x) \quad \text{or} \quad 30 = \left(\frac{1}{5}\right)(x)$$

$$\frac{30}{0.20} = \frac{0.20(x)}{0.20} \quad \text{or} \quad \left(\frac{5}{1}\right)\left(\frac{30}{1}\right) = \left(\frac{5}{1}\right)\left(\frac{1}{5}\right)(x)$$

$$\frac{30}{0.20} = \frac{\overset{1}{\cancel{0.20}}}{\underset{1}{\cancel{0.20}}}(x) \quad \text{or} \quad \left(\frac{5}{1}\right)\left(\frac{30}{1}\right) = \left(\frac{\overset{1}{\cancel{5}}}{1}\right)\left(\frac{1}{\underset{1}{\cancel{5}}}\right)(x)$$

$$150 = x \quad \text{or} \quad 150 = x$$

So 30 is 20% of 150.

(b) 40 is 80% of what number?

$$40 = 0.80(x) \quad \text{or} \quad 40 = \left(\frac{4}{5}\right)(x)$$

$$\frac{40}{0.80} = \frac{0.80(x)}{0.80} \quad \text{or} \quad \left(\frac{5}{4}\right)\left(\frac{40}{1}\right) = \left(\frac{5}{4}\right)\left(\frac{4}{5}\right)(x)$$

$$\frac{40}{0.80} = \frac{\overset{1}{\cancel{0.80}}}{\underset{1}{\cancel{0.80}}}(x) \quad \text{or} \quad \left(\frac{5}{\underset{1}{\cancel{4}}}\right)\left(\frac{\overset{10}{\cancel{40}}}{1}\right) = \left(\frac{\overset{1}{\cancel{5}}}{\underset{1}{\cancel{4}}}\right)\left(\frac{\overset{1}{\cancel{4}}}{\underset{1}{\cancel{5}}}\right)(x)$$

$$50 = x \quad \text{or} \quad 50 = x$$

So 40 is 80% of 50.

Percent–proportion method

Another simple method commonly used to solve any of the three types of percent problems is the **proportion** method (also called the *is/of method*).

First set up a blank proportion and then fill in the empty spaces by using the following steps.

$$\frac{\text{\%-number}}{100} = \frac{\text{"is"-number}}{\text{"of"-number}}$$

1. Whatever is next to the percent (%) is put over 100. (The word *what* is the unknown, or x.)
2. Whatever comes immediately after the word *of* goes on the bottom of one side of the proportion.
3. Whatever is left (comes next to the word *is*) goes on top, on one side of the proportion.
4. Then solve the problem.

Example 10: 30 is what percent of 50?

Set up a blank proportion.

$$\frac{\text{\%-number}}{100} = \frac{\text{"is"-number}}{\text{"of"-number}}$$

30 is what percent of 50?

Step 1: $\frac{x}{100} = \frac{\text{"is"-number}}{\text{"of"-number}}$

Step 2: $\frac{x}{100} = \frac{\text{"is"-number}}{50}$

Step 3: $\frac{x}{100} = \frac{30}{50}$

Step 4: $\frac{x}{100} = \frac{30}{50} = \frac{60}{100} = 60\%$

In this particular problem, however, it can be observed quickly that $\frac{30}{50} = \frac{60}{100} = 60\%$, so solving mechanically as shown is not time effective.

The proportion method works for the three basic types of percent questions:

- 40 is what percent of 200?

  (which is the same as what percent of 200 is 40?)

- 60 is 20% of what number?

  (which is the same as 20% of what number is 60?)

- What number is 15% of 30? (In this type, simply multiplying the numbers is probably easier.)

  (which is the same as 15% of 30 is what number?)

Example 11: Solve using the proportion method.

(a) 40 is what percent of 200?
(b) 60 is 20% of what number?
(c) What number is 15% of 30?

(a) 40 is what percent of 200?

$$\frac{x}{100} = \frac{40}{200}$$

$$200x = 4000$$

$$\frac{\overset{1}{\cancel{200}}}{\underset{1}{\cancel{200}}}\,x = \frac{\overset{20}{\cancel{4000}}}{\underset{1}{\cancel{200}}}$$

$$x = 20$$

So 40 is 20% of 200.

(b) 60 is 20% of what number?

$$\frac{20}{100} = \frac{60}{x}$$

$$20x = 6000$$

$$\frac{\overset{1}{\cancel{20}}}{\cancel{20}}\,x = \frac{\overset{300}{\cancel{6000}}}{\underset{1}{\cancel{20}}}$$

$$x = 300$$

So 60 is 20% of 300.

(c) What number is 15% of 30?

$$\frac{15}{100} = \frac{x}{30}$$

$$450 = 100x$$

$$\frac{450}{100} = \frac{\overset{1}{\cancel{100}}\,x}{\underset{1}{\cancel{100}}}$$

$$4.5 = x$$

So 4.5 is 15% of 30.

Finding percent increase or percent decrease

To find percent change (increase or decrease), use this formula:

$$\frac{\text{change}}{\text{starting amount}} \times 100\% = \text{percent change}$$

Example 12: Find the percent change.

(a) What is the percent decrease of a \$500 item on sale for \$400?

(a) change = 500 – 400 = 100

$$\frac{\text{change}}{\text{starting amount}} \times 100\% = \frac{\overset{1}{\cancel{100}}}{\underset{\overset{\cancel{5}}{1}}{\cancel{500}}} \times \frac{\overset{20}{\cancel{100}}}{1}\% = 20\% \text{ decrease}$$

(b) What is the percent increase of Jon's salary if it went from \$150 a month to \$200 a month?

(b) change = 200 – 150 = 50

$$\frac{\text{change}}{\text{starting amount}} \times 100\% = \frac{\overset{1}{\cancel{50}}}{\underset{3}{\cancel{150}}} \times \frac{100}{1}\% = \frac{100}{3}\% = 33\frac{1}{3}\% \text{ increase}$$

(c) What is the percent change from 2,100 to 1,890?

(c) change = 2,100 – 1,890 = 210

$$\frac{\text{change}}{\text{starting amount}} \times 100\% = \frac{\overset{1}{\cancel{210}}}{\underset{\underset{1}{\cancel{10}}}{\cancel{2100}}} \times \frac{\overset{10}{\cancel{100}}}{1}\% = 10\% \text{ change}$$

Note: The terms *percentage rise, percentage difference,* and *percentage change* are the same as *percent change.*

Chapter Check-Out

1. Change 0.04 to a percent.
2. Change 26% to a decimal.
3. Change $\frac{3}{20}$ to a percent.
4. Change 60% to a fraction.
5. What is 30% of 60?
6. 40 is what percent of 200?
7. 15 is 20% of what number?
8. What is the percent increase of a $400 item that is marked up to $500?

Answers: 1. 4% **2.** 0.26 **3.** 15% **4.** $\frac{3}{5}$ **5.** 18 **6.** 20% **7.** 75 **8.** 25%

Chapter 6

INTEGERS AND RATIONALS

Chapter Check-In

- ❑ Operations with integers
- ❑ Absolute value
- ❑ Operations with rational numbers
- ❑ Canceling

Integers

The term **integers** refers to all the whole numbers together with their opposites—not fractions or decimals.

For example, . . . –3, –2, –1, 0, 1, 2, 3, . . . are integers.

Number lines

On a number line, numbers to the right of 0 are positive. Numbers to the left of 0 are negative, as shown in Figure 6-1.

Figure 6-1 Number line showing integers.

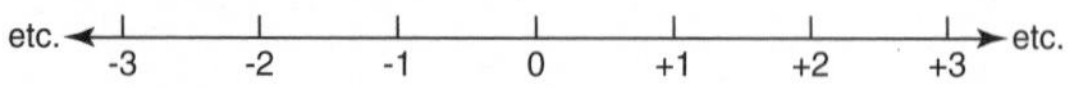

etc. -3 -2 -1 0 +1 +2 +3 etc.

This figure shows only the integers on the number line.

Given any two numbers on a number line, the one on the right is always larger, regardless of its sign (positive or negative).

Addition of integers

When adding two integers with the same sign (either both positive or both negative), add the integers and keep the same sign.

Example 1: Add the following.

(a) $\begin{array}{r} +3 \\ \underline{+\ +5} \end{array}$ (c) $\begin{array}{r} +4 \\ \underline{+\ +12} \end{array}$

(b) $\begin{array}{r} -6 \\ \underline{+\ -3} \end{array}$ (d) $\begin{array}{r} -8 \\ \underline{+\ -9} \end{array}$

(a) $\begin{array}{r} +3 \\ \underline{+\ +5} \\ +8 \end{array}$ (c) $\begin{array}{r} +4 \\ \underline{+\ +12} \\ +16 \end{array}$

(b) $\begin{array}{r} -6 \\ \underline{+\ -3} \\ -9 \end{array}$ (d) $\begin{array}{r} -8 \\ \underline{+\ -9} \\ -17 \end{array}$

When adding two integers with different signs (one positive and one negative), subtract the integers and keep the sign on the one with the larger value.

Example 2: Add the following.

(a) $\begin{array}{r} +8 \\ \underline{+\ -9} \end{array}$ (b) $\begin{array}{r} -30 \\ \underline{+\ +45} \end{array}$

(a) $\begin{array}{r} +8 \\ \underline{+\ -9} \\ -1 \end{array}$ (b) $\begin{array}{r} -30 \\ \underline{+\ +45} \\ +15 \end{array}$

Integers may also be added "horizontally."

Example 3: Add the following.

(a) $+8 + 11$ **(b)** $-15 + 7$ **(c)** $5 + (-3)$ **(d)** $-21 + 6$

(a) $+8 + 11 = +19$

(b) $-15 + 7 = -8$

(c) $5 + (-3) = +2$

(d) $-21 + 6 = -15$

Subtraction of integers

To subtract positive and/or negative integers, just change the sign of the number being subtracted and then use the rules for adding integers.

Example 4: Subtract the following.

(a) $\begin{array}{r} +12 \\ \underline{-\ \ +4} \end{array}$ **(c)** $\begin{array}{r} -19 \\ \underline{-\ \ +6} \end{array}$

(b) $\begin{array}{r} -14 \\ \underline{-\ \ -4} \end{array}$ **(d)** $\begin{array}{r} +20 \\ \underline{-\ \ -3} \end{array}$

(a) $\begin{array}{rr} 12 = & +12 \\ \underline{-\ \ +4} = & \underline{+\ -4} \\ & +8 \end{array}$ **(c)** $\begin{array}{rr} -19 = & -19 \\ \underline{-\ \ +6} = & \underline{+\ \ -6} \\ & -25 \end{array}$

(b) $\begin{array}{rr} -14 = & -14 \\ \underline{-\ \ -4} = & \underline{+\ \ +4} \\ & -10 \end{array}$ **(d)** $\begin{array}{rr} +20 = & +20 \\ \underline{-\ \ -3} = & \underline{+\ \ +3} \\ & +23 \end{array}$

Subtracting positive and/or negative integers may also be done "horizontally."

Example 5: Subtract the following.

(a) $+12 - (+4)$ **(c)** $-20 - (+3)$

(b) $+16 - (-6)$ **(d)** $-5 - (-2)$

(a) $+12 - (+4) = +12 + (-4) = 8$

(b) $+16 - (-6) = +16 + (+6) = 22$

(c) $-20 - (+3) = -20 + (-3) = -23$

(d) $-5 - (-2) = -5 + (+2) = -3$

Minus preceding a parenthesis

If a minus precedes a parenthesis, it means that everything within the parentheses is to be subtracted. Therefore, using the same rule as in subtraction of integers, simply change every sign within the parentheses to its opposite and then add.

Example 6: Subtract the following.

(a) $9 - (+3 - 5 + 7 - 6)$ **(b)** $20 - (+35 - 50 + 100)$

(a) $9 - (+3 - 5 + 7 - 6) = 9 + (-3 + 5 - 7 + 6)$

$= 9 + (+1)$

$= 10$

(b) $20 - (+35 - 50 + 100) = 20 + (-35 + 50 - 100)$

$= 20 + (-85)$

$= -65$

Or, if you can, total the numbers within the parentheses by first adding the positive numbers together, next adding the negative numbers together, then combining, and finally subtracting.

Example 7: Subtract the following.

(a) $9 - (+3 - 5 + 7 - 6)$

(b) $20 - (+35 - 50 + 100)$

(c) $3 - (1 - 4)$

(a) $9 - (+3 - 5 + 7 - 6) = 9 - (+10 - 11)$

$= 9 - (-1)$

$= 9 + (+1)$

$= 10$

(b) $20 - (+35 - 50 + 100) = 20 - (135 - 50)$

$= 20 - (85)$

$= -65$

(c) ***Remember:*** If there is no sign given, the number is understood to be positive.

$3 - (1 - 4) = 3 - (+1 - 4)$

$= 3 + (-1 + 4)$

$= 3 + 3$

$= 6$

or

$$\begin{aligned}3-(1-4)&=3-(-3)\\&=3+3\\&=6\end{aligned}$$

Multiplying and dividing integers

To multiply or divide integers, treat them just like regular numbers but remember this rule: An odd number of negative signs produces a negative answer. An even number of negative signs produces a positive answer.

Example 8: Multiply or divide the following.

(a) $(-4)\times(-7)$

(b) $(-1)\times(+3)\times(-8)\times(-1)$

(c) $(+5)\times(-6)\times(-2)\times(+4)\times(+3)$

(d) $\frac{-16}{-4}$

(e) $\frac{-48}{+3}$

(f) $\frac{+28}{-7}$

(a) $(-4)\times(-7)=+28$

(b) $(-1)\times(+3)\times(-8)\times(-1)=-24$

(c) $(+5)\times(-6)\times(-2)\times(+4)\times(+3)=+720$

(d) $\frac{-16}{-4}=+4$

(e) $\frac{-48}{+3}=-16$

(f) $\frac{+28}{-7}=-4$

Absolute value

The numerical value when direction or sign is not considered is called the absolute value. The absolute value of a number is written $|\ |$. Therefore, $|3| = 3$ and $|-4| = 4$. The absolute value of a number is always positive except when the number is 0. The absolute value of zero is zero, $|0| = 0$.

Example 9: Give the value.

(a) $|5|$ **(b)** $|-8|$ **(c)** $|3-9|$ **(d)** $3-|-6|$

(a) $|5| = 5$
(b) $|-8| = 8$
(c) $|3-9| = |-6| = 6$
(d) $3-|-6| = 3-6 = -3$

Note: Absolute value is taken first or work is done within absolute value brackets.

Rationals (Signed Numbers Including Fractions)

Recall that integers are positive and negative whole numbers and zero. When fractions and terminating or repeating decimals between the integers are included, the complete group of numbers is referred to as **rational numbers.** They are signed numbers including fractions. A more technical definition of a rational number is any number that can be written as a fraction with the numerator being a whole number or integer and the denominator being a natural number. (See Chapter 1.) Notice that fractions can be placed on the number line, as shown in Figure 6-2.

Figure 6-2 Number line showing integers and fractions.

etc. -3 $-2\frac{1}{2}$ -2 $-1\frac{1}{2}$ -1 $-\frac{1}{2}$ 0 $+\frac{1}{2}$ +1 $+1\frac{1}{2}$ +2 $+2\frac{1}{2}$ +3 etc.

Negative fractions

Fractions may be negative as well as positive. (See Chapter 4 for more information on fractions, as well as positive fractions.) Negative fractions are typically written as follows:

$$-\frac{3}{4} = \frac{-3}{4} = \frac{3}{-4}$$

Although they are all equal.

Adding positive and negative fractions

The rules for signs when adding integers applies to fractions as well. ***Remember:*** To add fractions, you must first get a common denominator.

Example 10: Add the following.

(a) $-\frac{1}{5}+\frac{3}{5}$ **(b)** $-\frac{2}{3}+\left(-\frac{1}{4}\right)$ **(c)** $-\frac{1}{2}+\frac{3}{8}$

(a) $-\frac{1}{5}+\frac{3}{5}=\frac{2}{5}$

(b) $-\frac{2}{3}+\left(-\frac{1}{4}\right)\rightarrow$

$$\begin{array}{rcr} -\frac{2}{3} & = & -\frac{8}{12} \\ +-\frac{1}{4} & = & +-\frac{3}{12} \\ \hline & & -\frac{11}{12} \end{array}$$

(c) $-\frac{1}{2}+\frac{3}{8}=-\frac{4}{8}+\frac{3}{8}=-\frac{1}{8}$

Adding positive and negative mixed numbers

The rules for signs when adding integers applies to mixed numbers as well.

Example 11: Add the following.

(a) $-2\frac{3}{4}+\left(-5\frac{1}{5}\right)$ (b) $-4\frac{1}{2}+2\frac{1}{3}$

(a) $-2\frac{3}{4}+\left(-5\frac{1}{5}\right)\rightarrow$

$$\begin{array}{rcr} -2\frac{3}{4} & = & -2\frac{15}{20} \\ +-5\frac{1}{5} & = & +-5\frac{4}{20} \\ \hline & & -7\frac{19}{20} \end{array}$$

(b) $-4\frac{1}{2}+2\frac{1}{3}=-4\frac{3}{6}+2\frac{2}{6}=-2\frac{1}{6}$

Subtracting positive and negative fractions

The rules for signs when subtracting integers applies to fractions as well. ***Remember:*** To subtract fractions, you must first get a common denominator.

Example 12: Subtract the following.

(a) $+\frac{7}{10}-\left(-\frac{1}{5}\right)$ (b) $+\frac{2}{7}-\left(-\frac{1}{4}\right)$ (c) $+\frac{1}{6}-\frac{3}{4}$

$$\text{(a)}\quad +\frac{7}{10}-\left(-\frac{1}{5}\right)=+\frac{7}{10}+\left(+\frac{1}{5}\right)\rightarrow \quad \begin{array}{rcr} +\dfrac{7}{10} & = & +\dfrac{7}{10} \\ \underline{+\ +\dfrac{1}{5}} & = & \underline{+\ +\dfrac{2}{10}} \\ & & +\dfrac{9}{10} \end{array}$$

$$\text{(b)}\quad +\frac{2}{7}-\left(-\frac{1}{4}\right)=+\frac{2}{7}+\left(+\frac{1}{4}\right)=+\frac{8}{28}+\left(+\frac{7}{28}\right)=+\frac{15}{28}$$

$$\text{(c)}\quad +\frac{1}{6}-\frac{3}{4}=+\frac{1}{6}+\left(-\frac{3}{4}\right)=+\frac{2}{12}+\left(-\frac{9}{12}\right)=-\frac{7}{12}$$

Subtracting positive and negative mixed numbers

The rules for signs when subtracting integers applies to mixed numbers as well. ***Remember:*** To subtract mixed numbers, you must first get a common denominator. If borrowing from a column is necessary, be cautious of simple mistakes.

Example 13: Subtract the following.

$$\text{(a)}\ +2\frac{1}{4}-\left(+3\frac{1}{2}\right) \qquad \text{(b)}\ -5\frac{1}{8}-\left(-2\frac{1}{3}\right) \qquad \text{(c)}\ +4\frac{2}{7}-7\frac{2}{5}$$

$$\text{(a)}\quad \begin{array}{rcrcr} +2\dfrac{1}{4} & = & +2\dfrac{1}{4} & \text{same as} & -3\dfrac{2}{4} \\ \underline{-\ +3\dfrac{1}{2}} & = & \underline{+\ -3\dfrac{2}{4}} & & \underline{+\ 2\dfrac{1}{4}} \\ & & & & -1\dfrac{1}{4} \end{array}$$

$$\text{(b)}\quad \begin{array}{rcrcrcr} -5\dfrac{1}{8} & = & -5\dfrac{1}{8} & = & -5\dfrac{3}{24} & = & -4\dfrac{27}{24} \\ \underline{-\ -2\dfrac{1}{3}} & = & \underline{+2\dfrac{1}{3}} & = & \underline{+2\dfrac{8}{24}} & = & \underline{+2\dfrac{8}{24}} \\ & & & & & & -2\dfrac{19}{24} \end{array}$$

$$\text{(c)}\quad +4\frac{2}{7}-7\frac{2}{5}=+4\frac{10}{35}-7\frac{14}{35}=-3\frac{4}{35}$$

Problems, such as the preceding ones, are usually most easily done by stacking the number with the larger absolute value on top, subtracting, and keeping the sign of the number with the larger absolute value.

Multiplying positive and negative fractions

The rules for signs when multiplying integers applies to fractions as well. ***Remember:*** To multiply fractions, multiply the numerators and then multiply the denominators. Always simplify to lowest terms if possible.

Example 14: Multiply the following.

(a) $\left(-\frac{2}{3}\right)\times\left(+\frac{1}{5}\right)$ **(b)** $\left(-\frac{5}{6}\right)\times\left(-\frac{7}{9}\right)$ **(c)** $\left(+\frac{3}{8}\right)\times\left(-\frac{2}{7}\right)$

(a) $\left(-\frac{2}{3}\right)\times\left(+\frac{1}{5}\right)=-\frac{2}{15}$

(b) $\left(-\frac{5}{6}\right)\times\left(-\frac{7}{9}\right)=+\frac{35}{54}$

(c) $\left(+\frac{3}{8}\right)\times\left(-\frac{2}{7}\right)=-\frac{6}{56}=-\frac{3}{28}$

Canceling

You can **cancel** when multiplying positive and negative fractions. Simply cancel as you do when multiplying positive fractions, but pay special attention to the signs involved. Follow the rules for signs when multiplying integers to obtain the proper sign. ***Remember:*** No sign means that a positive sign is understood.

Example 15: Multiply the following.

(a) $\left(-\frac{2}{3}\right)\times\left(\frac{5}{12}\right)$ **(b)** $\left(-\frac{5}{6}\right)\times\left(-\frac{3}{10}\right)$ **(c)** $\left(\frac{1}{5}\right)\times\left(-\frac{15}{16}\right)$

(a) $\left(-\frac{2}{3}\right)\times\left(\frac{5}{12}\right)=\left(-\frac{\overset{1}{\cancel{2}}}{3}\right)\times\left(\frac{5}{\underset{6}{\cancel{12}}}\right)=-\frac{5}{18}$

(b) $\left(-\frac{5}{6}\right)\times\left(-\frac{3}{10}\right)=\left(-\frac{\overset{1}{\cancel{5}}}{\underset{2}{\cancel{6}}}\right)\times\left(-\frac{\overset{1}{\cancel{3}}}{\underset{2}{\cancel{10}}}\right)=+\frac{1}{4}$

$$\text{(c)}\ \left(\frac{1}{5}\right)\times\left(-\frac{15}{16}\right)=\left(\frac{1}{\cancel{5}_1}\right)\times\left(-\frac{\cancel{15}^3}{16}\right)=-\frac{3}{16}$$

Multiplying positive and negative mixed numbers

Follow the rules for signs when multiplying integers to get the proper sign. ***Remember:*** Before multiplying mixed numbers, you must first change them to improper fractions.

Example 16: Multiply the following.

$$\text{(a)}\ \left(-3\frac{1}{4}\right)\times\left(2\frac{1}{2}\right)\quad \text{(b)}\ \left(6\frac{1}{2}\right)\times\left(-1\frac{1}{6}\right)\quad \text{(c)}\ \left(-4\frac{1}{4}\right)\times\left(-1\frac{1}{3}\right)$$

$$\text{(a)}\ \left(-3\frac{1}{4}\right)\times\left(2\frac{1}{2}\right)=\left(-\frac{13}{4}\right)\times\left(\frac{5}{2}\right)=-\frac{65}{8}=-8\frac{1}{8}$$

$$\text{(b)}\ \left(6\frac{1}{2}\right)\times\left(-1\frac{1}{6}\right)=\left(\frac{13}{2}\right)\times\left(-\frac{7}{6}\right)=-\frac{91}{12}=-7\frac{7}{12}$$

$$\text{(c)}\ \left(-4\frac{1}{4}\right)\times\left(-1\frac{1}{3}\right)=\left(-\frac{17}{4}\right)\times\left(-\frac{4}{3}\right)=\left(-\frac{17}{\cancel{4}_1}\right)\times\left(-\frac{\cancel{4}^1}{3}\right)=\frac{17}{3}=5\frac{2}{3}$$

Dividing positive and negative fractions

Follow the rules for signs when dividing integers to get the proper sign. ***Remember:*** When dividing fractions, first invert the divisor and then multiply.

Example 17: Divide the following.

$$\text{(a)}\ \left(-\frac{2}{3}\right)\div\left(\frac{1}{4}\right)\quad \text{(b)}\ \left(-\frac{2}{5}\right)\div\left(-\frac{3}{4}\right)\quad \text{(c)}\ \left(\frac{1}{6}\right)\div\left(-\frac{2}{3}\right)$$

$$\text{(a)}\ \left(-\frac{2}{3}\right)\div\left(\frac{1}{4}\right)=\left(-\frac{2}{3}\right)\times\left(\frac{4}{1}\right)=-\frac{8}{3}=-2\frac{2}{3}$$

$$\text{(b)}\ \left(-\frac{2}{5}\right)\div\left(-\frac{3}{4}\right)=\left(-\frac{2}{5}\right)\times\left(-\frac{4}{3}\right)=\frac{8}{15}$$

(c) $\left(\frac{1}{6}\right)\div\left(-\frac{2}{3}\right)=\left(\frac{1}{6}\right)\times\left(-\frac{3}{2}\right)=\left(\frac{1}{\cancel{6}_2}\right)\times\left(-\frac{\cancel{3}^1}{2}\right)=-\frac{1}{4}$

Dividing positive and negative mixed numbers

Follow the rules for signs when dividing integers to get the proper sign. ***Remember:*** Before dividing mixed numbers, you must first change them to improper fractions. Then you must invert the divisor and multiply.

Example 18: Divide the following.

(a) $\left(-2\frac{1}{2}\right)\div\left(\frac{1}{3}\right)$ **(b)** $\left(-3\frac{1}{2}\right)\div\left(-4\frac{1}{3}\right)$ **(c)** $\left(5\frac{1}{4}\right)\div\left(-3\frac{3}{5}\right)$

(a) $\left(-2\frac{1}{2}\right)\div\left(\frac{1}{3}\right)=\left(-\frac{5}{2}\right)\div\left(\frac{1}{3}\right)=\left(-\frac{5}{2}\right)\times\left(\frac{3}{1}\right)=-\frac{15}{2}=-7\frac{1}{2}$

(b) $\left(-3\frac{1}{2}\right)\div\left(-4\frac{1}{3}\right)=\left(-\frac{7}{2}\right)\div\left(-\frac{13}{3}\right)=\left(-\frac{7}{2}\right)\times\left(-\frac{3}{13}\right)=\frac{21}{26}$

(c)
$$\begin{aligned}\left(5\frac{1}{4}\right)\div\left(-3\frac{3}{5}\right)&=\left(\frac{21}{4}\right)\div\left(-\frac{18}{5}\right)\\&=\left(\frac{21}{4}\right)\times\left(-\frac{5}{18}\right)\\&=\left(\frac{\cancel{21}^7}{4}\right)\times\left(-\frac{5}{\cancel{18}_6}\right)\\&=-\frac{35}{24}\\&=-1\frac{11}{24}\end{aligned}$$

Chapter Check-Out

1. $-12 + 7 =$ _____.
2. $-13 - 17 =$ ______.
3. $26 - (-4) =$ ______.
4. $14 - (+4 - 6 + 8 - 5 - 3) =$ ____.
5. $(-7) \times (-6) =$ ____.
6. $-\frac{28}{7} =$ ______.
7. $|-9| =$ ______.
8. $-\frac{1}{4} + \frac{3}{8} =$ ______.
9. $\left(-\frac{2}{5}\right) - \left(-\frac{5}{8}\right) =$ ______.
10. $4\frac{5}{6} - \left(-3\frac{1}{3}\right) =$ ______.
11. $\left(-\frac{1}{2}\right) \times \left(-\frac{3}{5}\right) =$ ______.
12. $\left(-3\frac{3}{5}\right) \times \left(4\frac{1}{2}\right) =$ ______.
13. $\left(-\frac{7}{8}\right) \div \left(-\frac{2}{3}\right) =$ ______.
14. $\left(-8\frac{1}{6}\right) \div \left(2\frac{1}{4}\right) =$ ______.

Answers: 1. –5 **2.** –30 **3.** 30 **4.** 16 **5.** 42 **6.** –4 **7.** 9 **8.** $\frac{1}{8}$ **9.** $\frac{9}{40}$ **10.** $8\frac{1}{6}$ **11.** $\frac{3}{10}$ **12.** $-\frac{81}{5}$ or $-16\frac{1}{5}$ **13.** $\frac{21}{16}$ or $1\frac{5}{16}$ **14.** $-\frac{98}{27}$ or $-3\frac{17}{27}$

Chapter 7

POWERS, EXPONENTS, AND ROOTS

Chapter Check-In

- ❑ Operations with powers and exponents
- ❑ Square roots and cube roots
- ❑ Simplifying and approximating square roots

Powers and Exponents

Before you begin working with powers and exponents, some basic definitions are necessary.

Exponents

An **exponent** is a positive or negative number or 0 placed above and to the right of a quantity. It expresses the power to which the quantity is to be raised or lowered. In 4^3, 3 is the exponent. It shows that 4 is to be used as a factor three times: $4 \times 4 \times 4$ (multiplied by itself twice). 4^3 is read as four to the *third power* (or *four cubed*).

$$2^4 = 2 \times 2 \times 2 \times 2 = 16$$

$$3^2 = 3 \times 3 = 9$$

$$5^3 = 5 \times 5 \times 5 = 125$$

Remember: $x^1 = x$ and $x^0 = 1$ when x is any number (other than 0).

$$2^1 = 2 \qquad 2^0 = 1$$

$$3^1 = 3 \qquad 3^0 = 1$$

$$4^1 = 4 \qquad 4^0 = 1$$

Negative exponents

If the exponent is negative, such as 4^{-2}, the number and exponent may be dropped under the number 1 in a fraction to remove the negative sign.

Example 1: Simplify the following by first removing the negative signs and then removing the exponents.

(a) 4^{-2} **(b)** 5^{-3} **(c)** 2^{-4} **(d)** 3^{-1}

$$\text{(a)}\quad 4^{-2} = \frac{1}{4^2} = \frac{1}{16}$$

$$\text{(b)}\quad 5^{-3} = \frac{1}{5^3} = \frac{1}{125}$$

$$\text{(c)}\quad 2^{-4} = \frac{1}{2^4} = \frac{1}{16}$$

$$\text{(d)}\quad 3^{-1} = \frac{1}{3^1} = \frac{1}{3}$$

Squares and cubes

Two specific types of powers should be noted: **squares** and **cubes.** To square a number, just multiply it by itself the way you'd find the area of a square (the exponent is 2). For example, 6 squared (written 6^2) is 6×6, or 36. 36 is called a perfect square (the square of a whole number). Following is a partial list of perfect squares:

$1^2 = 1$

$2^2 = 4$

$3^2 = 9$

$4^2 = 16$

$5^2 = 25$

$6^2 = 36$

$7^2 = 49$

$8^2 = 64$

$9^2 = 81$

$10^2 = 100$

$11^2 = 121$

$12^2 = 144$

To **cube** a number, just multiply it by itself twice the way you'd find the volume of a cube (the exponent is 3). For example, 5 cubed (written 5^3 is

$5 \times 5 \times 5$, or 125. 125 is called a perfect cube (the cube of a whole number). Following is a partial list of perfect cubes.

$$1^3 = 1 \times 1 \times 1 = 1$$
$$2^3 = 2 \times 2 \times 2 = 8$$
$$3^3 = 3 \times 3 \times 3 = 27$$
$$4^3 = 4 \times 4 \times 4 = 64$$
$$5^3 = 5 \times 5 \times 5 = 125$$
$$6^3 = 6 \times 6 \times 6 = 216$$
$$7^3 = 7 \times 7 \times 7 = 343$$

Operations with powers and exponents

To multiply two numbers with exponents, if the base numbers are the same, simply keep the base number and add the exponents.

Example 2: Multiply the following, leaving the answers with exponents.

(a) $2^3 \times 2^5$ **(b)** $3^2 \times 3^5$ **(c)** $5^4 \times 5^7$

(a) $2^3 \times 2^5 = 2^{(3+5)} = 2^8$

$$\underbrace{2^3}_{2\times2\times2} \times \underbrace{2^5}_{2\times2\times2\times2\times2} = \underbrace{2^8}_{2\times2\times2\times2\times2\times2\times2\times2}$$

(b) $3^2 \times 3^5 = 3^{(2+5)} = 3^7$

(c) $5^4 \times 5^7 = 5^{(4+7)} = 5^{11}$

To divide two numbers with exponents, if the base numbers are the same, simply keep the base number and subtract the second exponent from the first, or the exponent of the denominator from the exponent of the numerator.

Example 3: Divide the following, leaving the answers with exponents.

(a) $5^6 \div 5^2$ (b) $\frac{8^7}{8^3}$

(a) $5^6 \div 5^2 = 5^{(6-2)} = 5^4$ $\quad 5^6 \div 5^2 = \frac{5^6}{5^2} = \frac{\cancel{5}^1 \times \cancel{5}^1 \times 5 \times 5 \times 5 \times 5}{\cancel{5}_1 \times \cancel{5}_1} = 5^4$

(b) $\frac{8^7}{8^3} = 8^{(7-3)} = 8^4$

To multiply or divide numbers with exponents, if the base numbers are different, you must simplify each number with an exponent first and then perform the operation.

Example 4: Simplify and perform the operation indicated.

(a) $2^3 \times 3^2$ **(b)** $6^2 \div 2^3$

(a) $2^3 \times 3^2 = 8 \times 9 = 72$

(b) $6^2 \div 2^3 = 36 \div 8 = 4\frac{4}{8} = 4\frac{1}{2}$

For problems such as those in Example 4, some shortcuts are possible.

To add or subtract numbers with exponents, whether the base numbers are the same or different, you must simplify each number with an exponent first and then perform the indicated operation.

Example 5: Simplify and perform the operation indicated.

(a) $3^2 - 2^3$ **(b)** $5^2 + 3^3$ **(c)** $4^2 + 9^3$ **(d)** $2^3 - 2^2$

(a) $3^2 - 2^3 = 9 - 8 = 1$
(b) $5^2 + 3^3 = 25 + 27 = 52$
(c) $4^2 + 9^3 = 16 + 729 = 745$
(d) $2^3 - 2^2 = 8 - 4 = 4$

If a number with an exponent is taken to another power $(4^2)^3$, simply keep the original base number and multiply the exponents.

Example 6: Multiply the following and leave the answers with exponents.

(a) $(6^3)^2$ **(b)** $(3^2)^4$ **(c)** $(5^4)^3$
(a) $(6^3)^2 = 6^{(3 \times 2)} = 6^6$
(b) $(3^2)^4 = 3^{(2 \times 4)} = 3^8$
(c) $(5^4)^3 = 5^{(4 \times 3)} = 5^{12}$

Square Roots and Cube Roots

Note: Square roots and cube roots and operations with them are often included in algebra books.

Square roots

To find the square root of a number, you want to find some number that when multiplied by itself gives you the original number. In other words, to find the square root of 25, you want to find the number that when multiplied by itself gives you 25. The square root of 25, then, is 5. The symbol for the square root is $\sqrt{\ }$. Following is a partial list of perfect (whole number) square roots.

$$\sqrt{0}=0 \quad \sqrt{16}=4 \quad \sqrt{64}=8$$
$$\sqrt{1}=1 \quad \sqrt{25}=5 \quad \sqrt{81}=9$$
$$\sqrt{4}=2 \quad \sqrt{36}=6 \quad \sqrt{100}=10$$
$$\sqrt{9}=3 \quad \sqrt{49}=7$$

Note: If no sign (or a positive sign) is placed in front of the square root, the positive answer is required. No sign means that a positive is understood. Only if a negative sign is in front of the square root is the negative answer required. This notation is used in many texts, as well as this book. Therefore,

$$\sqrt{4}=2 \text{ and } -\sqrt{4}=-2$$

Cube roots

To find the cube root of a number, you want to find some number that when multiplied by itself twice gives you the original number. In other words, to find the cube root of 8, you want to find the number that when multiplied by itself twice gives you 8. The cube root of 8, then, is 2 because $2 \times 2 \times 2 = 8$. Notice that the symbol for cube root is the radical sign with a small three (called the index) above and to the left, $\sqrt[3]{\ }$. Other roots are similarly defined and identified by the index given. (In square root, an index of 2 is understood and is usually not written.) Following is a partial list of perfect (whole number) cube roots.

$$\sqrt[3]{0}=0 \quad \sqrt[3]{64}=4 \quad \sqrt[3]{512}=8$$
$$\sqrt[3]{1}=1 \quad \sqrt[3]{125}=5 \quad \sqrt[3]{729}=9$$
$$\sqrt[3]{8}=2 \quad \sqrt[3]{216}=6 \quad \sqrt[3]{1000}=10$$
$$\sqrt[3]{27}=3 \quad \sqrt[3]{343}=7$$

Approximating square roots

To find the square root of a number that is not a perfect square, it is necessary to find an approximate answer by using the procedure given in Example 7.

Example 7: Approximate $\sqrt{42}$.

$$\sqrt{42} \text{ is between } \sqrt{36} \text{ and } \sqrt{49}$$
$$\sqrt{36} < \sqrt{42} < \sqrt{49}$$
$$\sqrt{36} = 6 \text{ and } \sqrt{49} = 7$$
$$\text{Therefore,} \qquad 6 < \sqrt{42} < 7$$

Since 42 is almost halfway between 36 and 49, $\sqrt{42}$ is almost halfway between $\sqrt{36}$ and $\sqrt{49}$. So $\sqrt{42}$ is approximately 6.5. To check, multiply the following:

$6.5 \times 6.5 = 42.25$ or about 42.

Example 8: Approximate $\sqrt{71}$.

$$\sqrt{64} < \sqrt{71} < \sqrt{81}$$
$$8 < \sqrt{71} < 9$$

Since $\sqrt{71}$ is slightly closer to $\sqrt{64}$ than it is to $\sqrt{81}$,

$$8 < 8.4 < 9$$
$$\sqrt{71} \approx 8.4$$

Check the answer.

$$\begin{array}{r} 8.4 \\ \times 8.4 \\ \hline 336 \\ 672 \\ \hline 70.56 \approx 71 \end{array}$$

Example 9: Approximate $\sqrt{\frac{300}{15}}$.

First, perform the operation under the radical.

$$\sqrt{\frac{300}{15}} = \sqrt{20}$$
$$\sqrt{16} < \sqrt{20} < \sqrt{25}$$
$$4 < \sqrt{20} < 5$$

Since $\sqrt{20}$ is slightly closer to $\sqrt{16}$ than it is to $\sqrt{25}$,

$$4 < 4.4 < 5$$

$$\sqrt{\frac{300}{15}} \approx 4.4$$

Square roots of nonperfect squares can be approximated, looked up in tables, or found by using a calculator. You may want to keep these two in mind, because they are commonly used.

$$\sqrt{2} \approx 1.414 \quad \sqrt{3} \approx 1.732$$

Simplifying square roots

Sometimes you will have to *simplify* square roots, or write them in simplest form. In fractions, $\frac{2}{4}$ can be simplified to $\frac{1}{2}$. In square roots, $\sqrt{32}$ can be simplified to $4\sqrt{2}$.

There are two main methods to *simplify a square root.*

Method 1:

Factor the number under the $\sqrt{\ }$ into two factors, one of which is the largest possible perfect square. (Perfect squares are 1, 4, 9, 16, 25, 36, 49, and so on)

Method 2:

Completely factor the number under the $\sqrt{\ }$ into prime factors and then simplify by bringing out any factors that came in pairs.

Example 10: Simplify $\sqrt{32}$.

Method 1.

$\sqrt{32} = \sqrt{16 \times 2}$

$= \sqrt{16} \times \sqrt{2}$

Take the square root of the perfect square number

$= 4 \times \sqrt{2}$

Finally, write it as a single expression.

$4\sqrt{2}$

Method 2.

$\sqrt{32} = \sqrt{2 \times 16}$

$= \sqrt{2 \times 2 \times 8}$

$= \sqrt{2 \times 2 \times 2 \times 4}$

$= \sqrt{2 \times 2 \times 2 \times 2 \times 2}$

Rewrite with pairs under the radical

$= \sqrt{2 \times 2} \times \sqrt{2 \times 2} \times \sqrt{2}$

$= 2 \times 2 \times \sqrt{2}$

$= 4\sqrt{2}$

In example 10, the largest perfect square is easy to see, and Method 1 probably is a faster method.

Example 11. Simplify $\sqrt{2016}$.

Method 1.

$$\sqrt{2016} = \sqrt{144 \times 14}$$
$$= \sqrt{144} \times \sqrt{14}$$
$$= 12\sqrt{14}$$

Method 2.

$$\sqrt{2016} = \sqrt{2 \times 1008}$$
$$= \sqrt{2 \times 2 \times 504}$$
$$= \sqrt{2 \times 2 \times 2 \times 252}$$
$$= \sqrt{2 \times 2 \times 2 \times 2 \times 126}$$
$$= \sqrt{2 \times 2 \times 2 \times 2 \times 2 \times 63}$$
$$= \sqrt{2 \times 2 \times 2 \times 2 \times 2 \times 3 \times 21}$$
$$= \sqrt{2 \times 2 \times 2 \times 2 \times 2 \times 3 \times 3 \times 7}$$
$$= \sqrt{2 \times 2} \times \sqrt{2 \times 2} \times \sqrt{3 \times 3} \times \sqrt{2 \times 7}$$
$$= 2 \times 2 \times 3 \times \sqrt{14}$$
$$= 12\sqrt{14}$$

In example 11, it is not so obvious that the largest perfect square is 144, so method 2 is probably the faster method.

Example 12: Simplify $\sqrt{75}$.

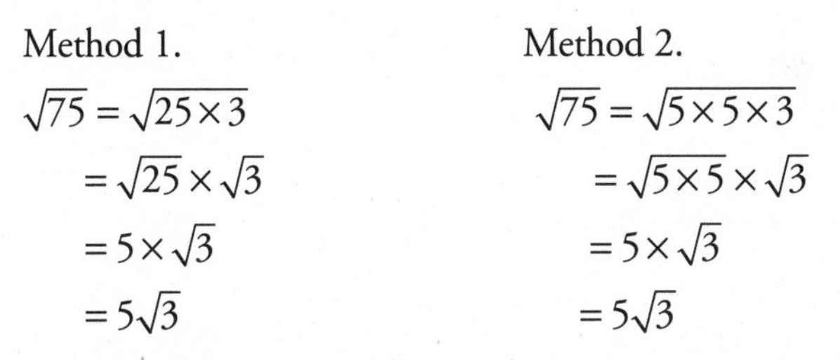

Method 1.

$$\sqrt{75} = \sqrt{25 \times 3}$$
$$= \sqrt{25} \times \sqrt{3}$$
$$= 5 \times \sqrt{3}$$
$$= 5\sqrt{3}$$

Method 2.

$$\sqrt{75} = \sqrt{5 \times 5 \times 3}$$
$$= \sqrt{5 \times 5} \times \sqrt{3}$$
$$= 5 \times \sqrt{3}$$
$$= 5\sqrt{3}$$

Remember: Most square roots cannot be simplified since they are already in simplest form, such as $\sqrt{7}, \sqrt{10}, \sqrt{15}$.

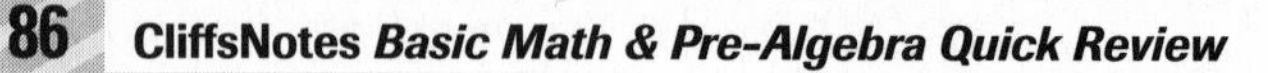

Chapter Check-Out

1. $5^4 =$ ______.
2. $6^{-2} =$ ______.
3. $3^2 \times 3^5 =$ ______. (with exponents)
4. $5^8 \div 5^3 =$ ______. (with exponents)
5. $4^2 - 3^3 =$ ______.
6. $(4^3)^2 =$ ______. (with exponents)
7. $\sqrt[3]{125} =$ ______.
8. Approximate $\sqrt{29}$ to the nearest tenth.
9. Simplify $\sqrt{72}$.

Answers: 1. 625 **2.** $\frac{1}{36}$ **3.** 3^7 **4.** 5^5 **5.** –11 **6.** 4^6 **7.** 5 **8.** 5.4 **9.** $6\sqrt{2}$

Chapter 8

POWERS OF TEN AND SCIENTIFIC NOTATION

Chapter Check-In

- ❑ Operations with powers of ten
- ❑ Changing to scientific notation
- ❑ Operations with scientific notation

Powers of Ten

Since our number system is based on **powers** of ten, you should understand the notation and how to work with these powers, as follows:

$$10^0 = 1$$
$$10^1 = 10$$
$$10^2 = 10 \times 10 = 100$$
$$10^3 = 10 \times 10 \times 10 = 1,000$$

and so on.

$$10^{-1} = \frac{1}{10^1} = \frac{1}{10} = 0.1$$
$$10^{-2} = \frac{1}{10^2} = \frac{1}{100} = 0.01$$
$$10^{-3} = \frac{1}{10^3} = \frac{1}{1000} = 0.001$$

and so on.

Multiplying powers of ten

To multiply powers of ten, add the exponents.

Example 1: Multiply the following and leave the answers in powers of ten.

(a) 100×10

(b) $1{,}000 \times 100$

(c) $0.01 \times .001$

(d) $10{,}000 \times 0.01$

(e) $0.0001 \times 1{,}000$

(a) $100 \times 10 = 10^2 \times 10^1 = 10^{(2+1)} = 10^3$

(b) $1{,}000 \times 100 = 10^3 \times 10^2 = 10^{(3+2)} = 10^5$

(c) $0.01 \times 0.001 = 10^{-2} \times 10^{-3} = 10^{[-2+(-3)]} = 10^{-5}$

(d) $10{,}000 \times 0.01 = 10^4 \times 10^{-2} = 10^{[4+(-2)]} = 10^2$

(e) $0.0001 \times 1{,}000 = 10^{-4} \times 10^3 = 10^{[-4+3]} = 10^{-1}$

Dividing powers of ten

To divide powers of 10, subtract the exponents; that is, subtract the exponent of the second number (the *divisor*) from the exponent of the first number (the dividend).

Example 2: Divide the following and leave the answers in powers of ten.

(a) $1{,}000 \div 100$

(b) $100 \div 10{,}000$

(c) $1 \div 0.01$

(d) $0.001 \div 0.01$

(e) $10{,}000 \div 0.1$

(a) $1{,}000 \div 100 = 10^3 \div 10^2 = 10^{(3-2)} = 10^1$ or 10

(b) $100 \div 10{,}000 = 10^2 \div 10^4 = 10^{(2-4)} = 10^{-2}$

(c) $1 \div 0.01 = 10^0 \div 10^{-2} = 10^{[0-(-2)]} = 10^{(0+2)} = 10^2$

(d) $0.001 \div 0.01 = 10^{-3} \div 10^{-2} = 10^{[-3-(-2)]} = 10^{(-3+2)} = 10^{-1}$

(e) $10{,}000 \div 0.1 = 10^4 \div 10^{-1} = 10^{[4-(-1)]} = 10^{(4+1)} = 10^5$

Scientific Notation

Very large or very small numbers are sometimes written in **scientific notation.** A number written in scientific notation is a number between 1 and 10 multiplied by a power of 10.

Example 3: Express the following in scientific notation.

(a) 3,400,000

(b) 0.0000008

(c) 0.0047

(d) 274.3

(a) 3,400,000 written in scientific notation is 3.4×10^6. Simply place the decimal point to get a number between 1 and 10 and then count the digits to the right of the decimal to get the power of 10.

3.400000.

moved six digits to the left

(b) 0.0000008 written in scientific notation is 8×10^{-7}.

Simply place the decimal point to get a number between 1 and 10 and then count the digits from the original decimal point to the new one.

.0000008.

moved seven digits to the right

(c) 0.0047 written in scientific notation is 4.7×10^{-3}.

Simply place the decimal point to get a number between 1 and 10 and then count the digits from the original decimal point to the new one.

.004.7

moved three digits to the right

(d) 274.3 written in scientific notation is 2.743×10^2.

Simply place the decimal point to get a number between 1 and 10 and then count the digits from the original point to the new one.

2.74.3

moved two digits to the left

Notice that numbers greater than 1 have positive exponents when expressed in scientific notation and numbers between 0 and 1 have negative exponents when expressed in scientific notation.

Multiplication in scientific notation

To multiply numbers in scientific notation, simply multiply the numbers that are between 1 and 10 together to get the first number and add the powers of ten to get the second number.

Example 4: Multiply the following and express the answers in scientific notation.

(a) $(2 \times 10^2)(3 \times 10^4)$

(b) $(6 \times 10^5)(5 \times 10^7)$

(c) $(4 \times 10^{-4})(2 \times 10^5)$

(d) $(5 \times 10^4)(9 \times 10^2)$

(e) $(2 \times 10^2)(4 \times 10^4)(5 \times 10^6)$

(a) $$(2 \times 10^2)(3 \times 10^4) = (2 \times 10^2)(3 \times 10^4) = 6 \times 10^6$$

(× the first numbers, + the exponents)

(b) $$(6 \times 10^5)(5 \times 10^7) = (6 \times 10^5)(5 \times 10^7) = 30 \times 10^{12}$$

(× the first numbers, + the exponents)

This answer must be changed to scientific notation.

Change the 30 into its scientific notation form.

$$30 \times 10^{12} = 3.0 \times 10^1 \times 10^{12} = 3.0 \times 10^{13}$$

(c) $$(4 \times 10^{-4})(2 \times 10^5) = (4 \times 10^{-4})(2 \times 10^5) = 8 \times 10^1$$

(× the first numbers, + the exponents)

(d) $(5 \times 10^4)(9 \times 10^2) = (5 \times 10^4)(9 \times 10^2)$ (× joins the numbers, + joins the exponents)

$$= 45 \times 10^6$$
$$= 4.5 \times 10^1 \times 10^6$$
$$= 4.5 \times 10^7$$

(e) $(2 \times 10^2)(4 \times 10^4)(5 \times 10^6) = (2 \times 10^2)(4 \times 10^4)(5 \times 10^6)$ (× joins the numbers, + joins the exponents)

$$= 40 \times 10^{12}$$
$$= 4.0 \times 10^1 \times 10^{12}$$
$$= 4.0 \times 10^{13}$$

Division in scientific notation

To divide numbers in scientific notation, simply divide the numbers that are between 1 and 10 to get the first number and subtract the powers of ten to get the second number.

Example 5: Divide the following and express the answers in scientific notation.

(a) $\left(8\times10^5\right)\div\left(2\times10^2\right)$ **(d)** $\left(2\times10^4\right)\div\left(5\times10^2\right)$

(b) $\dfrac{\left(7\times10^9\right)}{\left(4\times10^3\right)}$ **(e)** $\left(8.4\times10^5\right)\div\left(2.1\times10^{-4}\right)$

(c) $\left(6\times10^7\right)\div\left(3\times10^9\right)$

(a) $\left(8\times10^5\right)\div\left(2\times10^2\right)=(8\div2)\times\left(10^5\div10^2\right)$

$$=4\times10^{(5-2)}$$
$$=4\times10^3$$

(b) $\dfrac{\left(7\times10^9\right)}{\left(4\times10^3\right)}=(7\div4)\times\left(10^9\div10^3\right)$

$$=1.75\times10^{(9-3)}$$
$$=1.75\times10^6$$

(c) $(6 \times 10^7) \div (3 \times 10^9) = (6 \div 3) \times (10^7 \div 10^9)$

$= 2 \times 10^{(7-9)}$

$= 2 \times 10^{-7}$

(d) $(2 \times 10^4) \div (5 \times 10^2) = (2 \times 10^4) \div (5 \times 10^2)$ (arcs: $\div$, $-$)

$= 0.4 \times 10^{(4-2)}$

$= 0.4 \times 10^2$

Change 0.4 into its scientific notation form.

$= 4 \times 10^{-1} \times 10^2$

$= 4 \times 10^1$

(e) $(8.4 \times 10^5) \div (2.1 \times 10^{-4}) = (8.4 \times 10^5) \div (2.1 \times 10^{-4})$ (arcs: $\div$, $-$)

$= 4 \times 10^{5-(-4)}$

$= 4 \times 10^9$

Chapter Check-Out

1. Multiply $1{,}000 \times 100$ and leave the answer in powers of ten.
2. Multiply 0.1×0.001 and leave the answer in powers of ten.
3. Divide $100 \div 1{,}000$ and leave the answer in powers of ten.
4. Express 27,000 in scientific notation.
5. Express 0.0005 in scientific notation.
6. $(3 \times 10^4)(2 \times 10^3) =$ ______. (in scientific notation)
7. $(4 \times 10^{-7})(5 \times 10^2) =$ ______. (in scientific notation)
8. $(9 \times 10^4) \div (3 \times 10^{-2}) =$ ______. (in scientific notation)

Answers: 1. 10^5 **2.** 10^{-4} **3.** 10^{-1} **4.** 2.7×10^4 **5.** 5×10^{-4} **6.** 6×10^7 **7.** 2×10^{-4} **8.** 3×10^6

Chapter 9
MEASUREMENTS

Chapter Check-In

- ❑ U.S. customary and metric systems
- ❑ Converting units of measure
- ❑ Precision in measurement
- ❑ Significant digits
- ❑ Calculating perimeter and area of figures
- ❑ Calculating circumference and area of a circle

You should become familiar with the two measurement systems—the U.S. customary system and the metric system.

U.S. Customary System

The U.S. customary system of measurement is used throughout the United States, although the metric system is being phased in as well. You should be familiar with the following basic measurements of the U.S. customary system:

- **Length**

  12 inches (in) = 1 foot (ft)

  3 feet = 1 yard (yd)

  36 inches = 1 yard

  1,760 yards = 1 mile (mi)

  5,280 feet = 1 mile

- **Area**

  144 square inches (sq in) = 1 square foot (sq ft)

  9 square feet = 1 square yard (sq yd)

- **Weight**

  16 ounces (oz) = 1 pound (lb)

  2000 pounds = 1 ton (T)

- **Capacity**

  2 cups = 1 pint (pt)

  2 pints = 1 quart (qt)

  4 quarts = 1 gallon (gal)

  4 pecks = 1 bushel

- **Time**

  365 days = 1 year

  52 weeks = 1 year

  10 years = 1 decade

  100 years = 1 century

Metric System

The metric system of measurement is based on powers of ten. To understand the metric system, you should know the meaning of the prefixes to each base unit.

- Prefixes

  kilo = thousand

  hecto = hundred

  deka = ten

  deci = tenth

  centi = hundredth

  milli = thousandth

Now, applying these prefixes to length, volume, and mass should be easier.

- Length—meter

    1 kilometer (km) = 1,000 meters (m)

    1 hectometer (hm) = 100 meters

    1 dekameter (dkm) = 10 meters

    10 decimeters (dm) = 1 meter

    100 centimeters (cm) = 1 meter

    1,000 millimeters (mm) = 1 meter

- Volume—liter

    1,000 milliliters (ml, or mL) = 1 liter (l, or L)

    1,000 liters = 1 kiloliter (kl, or kL)

- Mass—gram

    1,000 milligrams (mg) = 1 gram (g)

    1,000 grams = 1 kilogram (kg)

    1,000 kilograms = 1 metric ton (t)

- Approximations

    A meter is a little more than a yard.

    A kilometer is about 0.6 mile.

    A kilogram is about 2.2 pounds.

    A liter is slightly more than a quart.

Converting Units of Measure

Example 1: If 36 inches equal 1 yard, 3 yards equal how many inches?

Intuitively, $3 \times 36 = 108$ in

By proportion, using yards over inches,

$$\frac{\text{yards}}{\text{inches}}: \ \frac{1}{36} = \frac{3}{x}$$

Remember: Place the same units across from each other—for example, inches across from inches. Then solve.

$$\frac{1}{36} = \frac{3}{x}$$

Cross multiply.

$$x = 108 \text{ in}$$

Example 2: If 2.2 pounds equal 1 kilogram, 10 pounds equal approximately how many kilograms?

Intuitively, $10 \div 2.2 \approx 4.5$ kg

By proportion, using kilograms over pounds,

$$\frac{\text{kilograms}}{\text{pounds}} : \frac{1}{2.2} = \frac{x}{10}$$

Cross multiply.

$$2.2x = 10$$

Divide and round off to the nearest tenth.

$$\frac{\overset{1}{\cancel{2.2}}x}{\underset{1}{\cancel{2.2}}} = \frac{10}{2.2}$$

$$x \approx 4.5 \text{ kg}$$

Example 3: Change 3 decades into weeks.

1 decade equals 10 years, so 3 decades equal 30 years.

1 year equals 52 weeks, so

$$\frac{30}{x} = \frac{1}{52}$$

$$30 \cancel{\text{ years}} \times 52 \frac{\text{weeks}}{\cancel{\text{year}}} = 1{,}560 \text{ weeks}$$

Therefore, there are 1,560 weeks in 3 decades.

Notice that this was converted step by step. It can be done in one step, as follows:

$$3\,\cancel{\text{decades}} \times 10\frac{\cancel{\text{years}}}{\cancel{\text{decade}}} \times 52\frac{\text{weeks}}{\cancel{\text{year}}} = 1{,}560 \text{ weeks}$$

$$3 \times 10 \times 52 = 1{,}560 \text{ weeks.}$$

Example 4: If 1,760 yards equal 1 mile, how many yards are in 5 miles?

$$\frac{1{,}760}{1} = \frac{x}{5}$$

$$1{,}760\,\frac{\text{yards}}{\cancel{\text{mile}}} \times 5\,\cancel{\text{miles}} = 8{,}800\,\text{yards}$$

Therefore there are 8,800 yards in 5 miles.

Example 5: If 1 kilometer equals 1,000 meters, and 1 dekameter equals 10 meters, how many dekameters are in 3 kilometers?

Since there are 1,000 meters in one kilometer, there will be 3,000 meters in 3 kilometers. Now use a proportion that compares dekameters with meters

$$\frac{\text{dekameter}}{\text{meter}}:\ \frac{1}{10} = \frac{x}{3{,}000}$$

$$10x = 3{,}000$$

$$\frac{\overset{1}{\cancel{10}}\,x}{\underset{1}{\cancel{10}}} = \frac{3{,}000}{10}$$

$$x = 300 \text{ dekameters}$$

Therefore there are 300 dekameters in 3 kilometers.

Precision

The word *precision* refers to the degree of exactness of a measurement, that is, how fine the measurement is. The smaller the unit of measure used, the more precise is the measurement. Keep in mind that the precision of a measurement has nothing to do with the size of the numbers, only with the unit used.

Example 6: Which of the following measurements is more precise?

(a) 5.4 mm or 3.22 mm

(b) 1 m or 1 km

(c) 3 ft or 11 in

(d) 3.67 m or 5.1 m

(e) 5.69 cm or 9.99 cm

(a) The smallest unit in 5.4 mm is 0.4, and the smallest unit in 3.22 mm is 0.02, so 3.22 mm is more precise.

(b) Meter (m) is a smaller unit than kilometer (km), so 1 m is more precise.

(c) Inch (in) is a smaller unit than foot (ft), so 11 inches is more precise.

(d) 3.67 goes to the hundredths, and 5.1 goes to the tenths, so 3.67 m is more precise.

(e) Each one goes to the hundredths place, so they have the same precision.

When adding or subtracting two measures, you cannot be more precise than the least precise unit being used. In other words, the unit being used in the answer should be the same as the less precise of the units used in the two measurements.

Example 7: Perform the indicated operation and give the answer in the less precise unit.

(a) 5.44 km + 2.1 km

(b) 32.77 g – 12 g

First round off the more precise number and then calculate.

(a)
$$\begin{array}{rcr} 5.44\text{ km} & \approx & 5.4\text{ km} \\ +2.1\text{ km} & = & 2.1\text{ km} \\ \hline & & 7.5\text{ km} \end{array}$$

(b)
$$\begin{array}{rcr} 32.77\text{g} & \approx & 33\text{g} \\ -12\text{g} & = & 12\text{g} \\ \hline & & 21\text{g} \end{array}$$

Significant Digits

When a digit tells how many units of measure are involved, it is a significant digit. To find the number of units of measure, simply divide the actual measurement by the unit of measure.

Example 8: Find the number of units used and the significant digits for each of the following.

(a) 23.7 m (unit of measure, 0.1 m)

(b) 0.05 cm (unit of measure, 0.01 cm)

(c) 520 g (unit of measure, 1 g)

(d) 760 km (unit of measure, 10 km)

(e) 2100 m (unit of measure, 10 m)

(f) 2100 m (unit of measure, 100 m)

For each of these examples, find the number of units of measure by dividing the actual measurement by the unit of measure. For example, for (a),

$$0.1\overline{)23.7} = 237$$

	Measurement	*Unit of Measure*	*Numbers of Units Used*	*Significant Digits*
(a)	23.7 m	0.1 m	237 units of 0.1 m	237
(b)	0.05 cm	0.01 cm	5 units of 0.01 cm	5
(c)	520 g	1 g	520 units of 1 g	520
(d)	760 km	10 km	76 units of 10 km	76
(e)	2100 m	10 m	210 units of 10m	210
(f)	2100 m	100 m	21 units of 100 m	21

The following rules can be used as general guidelines for determining significant digits:

- Digits that are not zero are always significant. For example, all three digits are significant in 35.7.
- Zeros at the end (or right) of the decimal are always significant. For example, both digits are significant in .20.

- In a decimal, zeros in front (or to the left) of a significant digit are never significant. For example, only 7 is significant in .007.
- All zeros that are between significant digits are significant. For example, all digits are significant in 500.5.
- Depending on the unit of measure, zeros that are at the end (or right) of a whole number may or may not be significant. For example, 500 m measured to the 1 m gives three significant digits (5, 0, and 0), whereas 500 m measured to the 10 m gives two significant digits (5 and 0).

When computing using significant digits, you should always round the answer to the smallest number of significant digits in any of the numbers being used.

Example 9: Find the area of the rectangle with length 3.1 m and width 2.2 m.

$$\begin{array}{rl} 3.1 & \text{(2 significant digits)} \\ \times\ 2.2 & \text{(2 significant digits)} \\ \hline 6.82 & \text{(3 significant digits must be changed to 2 significant digits)} \end{array}$$

So $6.82 \approx 6.8$ sq m

Example 10: Find the length of a rectangle with area 19 sq cm and width 4 cm.

Because 19 has two significant digits and 4 has one significant digit, the answer must be rounded to one significant digit.

$$4\overline{)19.00} = 4.75$$

$$\text{So } 4.75 \approx 5 \text{ cm.}$$

Calculating Measurements of Basic Figures

Some basic figures, such as squares, rectangles, parallelograms, trapezoids, triangles, and circles, have measurements that are not difficult to calculate if the necessary information is given and the proper formula is used. You should first be familiar with the formulas of these basic figures.

Perimeter of some polygons—squares, rectangles, parallelograms, trapezoids, and triangles

Perimeter (P) means the total distance all the way around the outside of the polygon (a many-sided plane closed figure). The perimeter of that polygon can be determined by adding up the lengths of all the sides. The total distance around is the sum of all sides of the polygon. No special formulas are necessary, although the following two formulas are commonly seen:

- Perimeter (P) of a square and a rhombus = $4s$ (s = length of side).
- Perimeter (P) of a parallelogram and a rectangle = $2l + 2w$ or $2(l + w)$ (l = length, w = width).

Area of polygons—squares, rectangles, parallelograms, trapezoids, and triangles

Area (A) means the amount of space inside the polygon. Each type of polygon has a formula to determine its area.

A triangle is a three-sided polygon. In a triangle, the base is the side the triangle is resting on, and the height is the distance from the base to the opposite point, or vertex.

$$\text{Triangle}: A = \frac{1}{2}bh \ (b = \text{base}, h = \text{height}). \text{ (See Figure 9-1.)}$$

Figure 9-1 Triangles showing base and height.

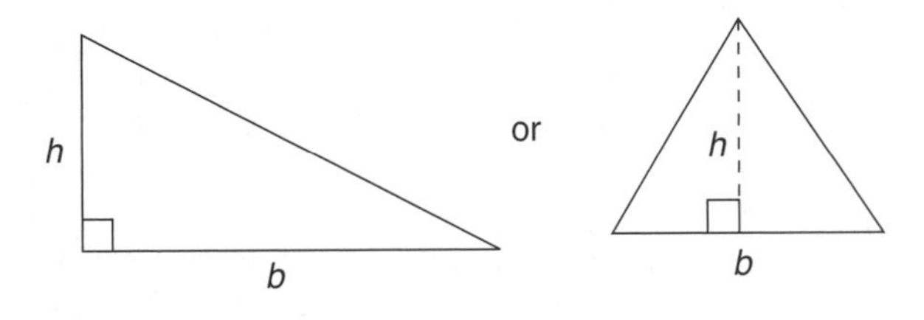

Example 11: What is the area of the triangle shown in Figure 9-2?

Figure 9-2 Triangle showing base and height.

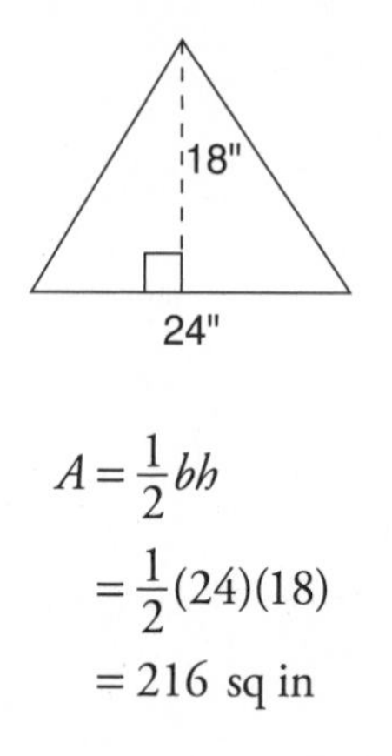

$$\begin{aligned} A &= \frac{1}{2}bh \\ &= \frac{1}{2}(24)(18) \\ &= 216 \text{ sq in} \end{aligned}$$

A square is a four-sided polygon with all sides equal and all right angles (90 degrees). A rectangle is a four-sided polygon with opposites sides equal and all right angles. In a square or rectangle, the bottom, or resting side, is the base, and either adjacent side is the height.

Square or rectangle: $A = lw$. (See Figure 9-3.)

Figure 9-3 Square and rectangle showing length and width.

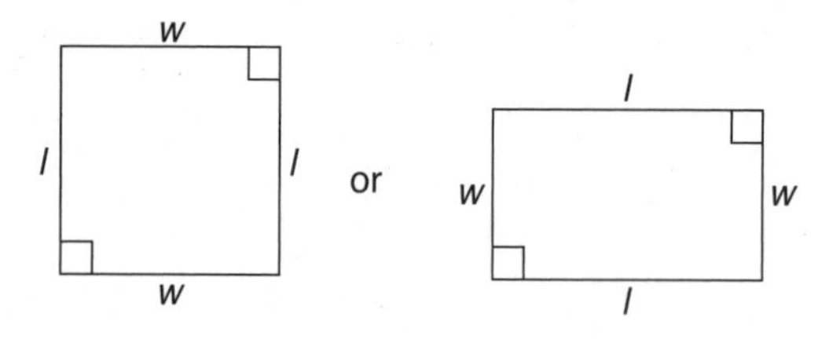

Example 12: What is the area of these polygons?

(a) The square shown in Figure 9-4(a)

(b) The rectangle shown in Figure 9-4(b)

Figure 9-4 Square and rectangle.

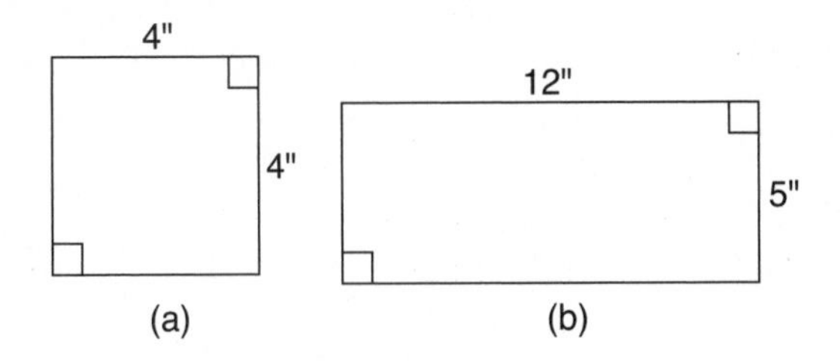

(a) $A = lw$
$= 4(4)$
$= 16$ sq in

(b) $A = lw$
$= 12(5)$
$= 60$ sq in

A parallelogram is a four-sided polygon with opposite sides parallel and equal. In a parallelogram, the resting side is usually considered the base, and a perpendicular line going from the base to the side opposite this base is the height.

Parallelogram: $A = bh$. (See Figure 9-5.)

Figure 9-5 Parallelogram showing base and height.

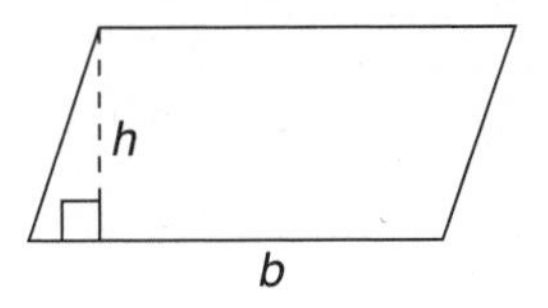

Example 13: What is the area of the parallelogram shown in Figure 9-6?

Figure 9-6 Parallelogram.

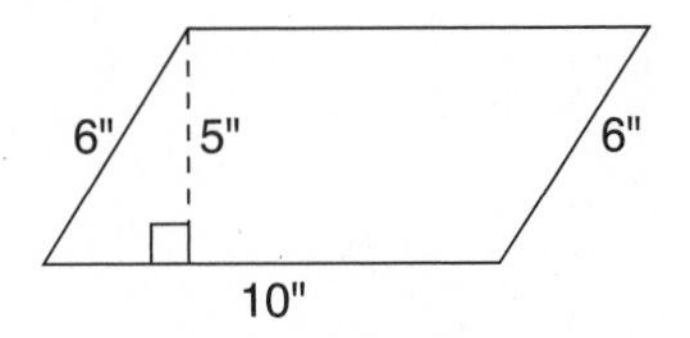

$$A = bh$$
$$= 10(5)$$
$$= 50 \text{ sq in}$$

A trapezoid is a four-sided polygon with only two sides parallel. In a trapezoid, the parallel sides are the bases, and the distance between the two bases is the height.

Trapezoid: $A = \frac{1}{2}(b_1 + b_2)h$. (See Figure 9-7.)

Figure 9-7 Trapezoid showing bases and height.

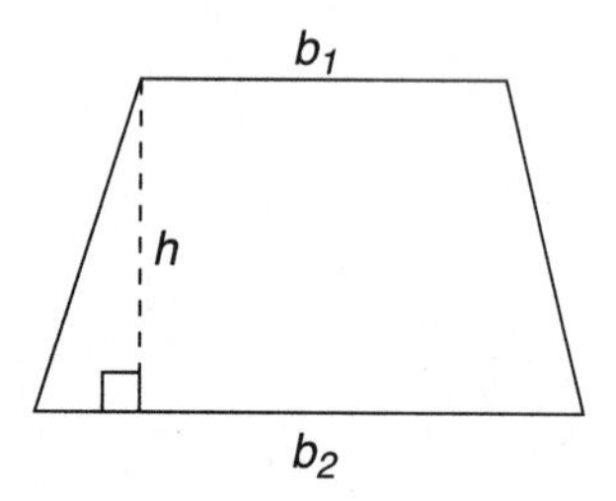

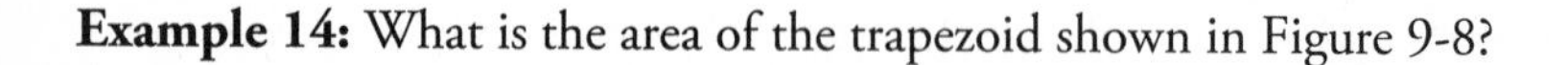

Example 14: What is the area of the trapezoid shown in Figure 9-8?

Figure 9-8 Trapezoid.

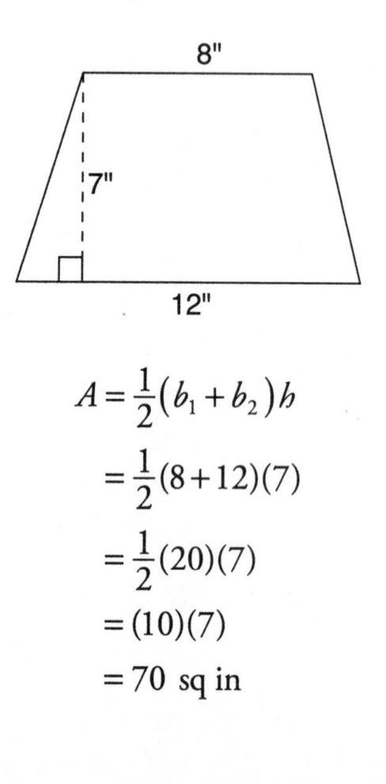

$$
\begin{aligned}
A &= \frac{1}{2}(b_1 + b_2)h \\
&= \frac{1}{2}(8+12)(7) \\
&= \frac{1}{2}(20)(7) \\
&= (10)(7) \\
&= 70 \text{ sq in}
\end{aligned}
$$

Example 15: What is the perimeter (*P*) and the area (*A*) of the polygons shown in Figure 9-9, (a) through (f), in which all measures are given in inches?

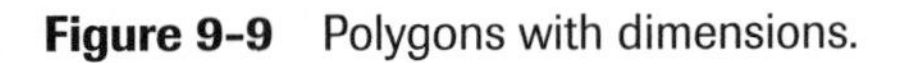

Figure 9-9 Polygons with dimensions.

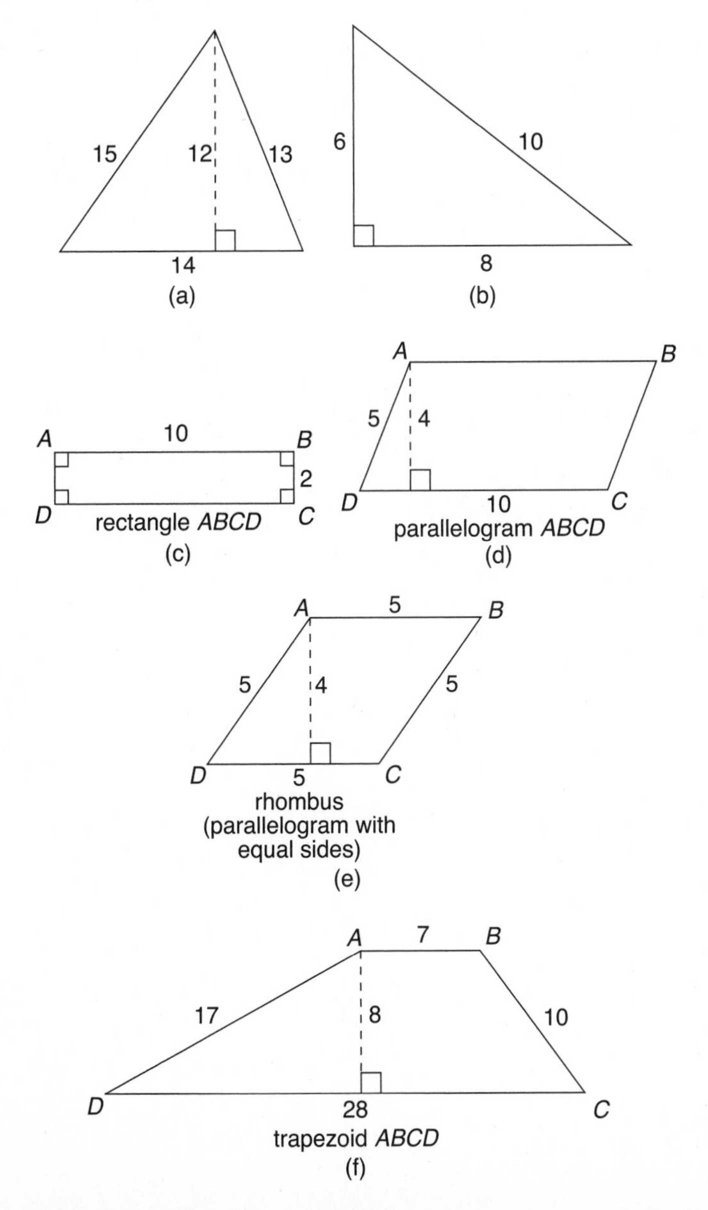

(a) $P = 15 + 13 + 14$
$= 42 \text{ in}$

$A = \frac{1}{2}bh$
$= \frac{1}{2}(14)(12)$
$= 84 \text{ sq in}$

(b) $P = 6 + 8 + 10$
$= 24 \text{ in}$

$A = \frac{1}{2}bh$
$= \frac{1}{2}(8)(6)$
$= \frac{1}{2}(48)$
$= 24 \text{ sq in}$

(c) $P = 2(10 + 2) = 10 + 10 + 2 + 2$
$= 24 \text{ in}$

$A = bh$
$= (10)(2)$
$= 20 \text{ sq in}$

(d) $P = 2(10 + 5) = 10 + 10 + 5 + 5$
$= 30 \text{ in}$

$A = bh$
$= (10)(4)$
$= 40 \text{ sq in}$

(e) $P = 4(5) = 5 + 5 + 5 + 5$
$= 20 \text{ in}$

$A = bh$
$= (4)(5)$
$= 20 \text{ sq in}$

(f) $P = 17 + 7 + 10 + 28$
$= 62 \text{ in}$

$A = \frac{1}{2}(b_1 + b_2)h$
$= \frac{1}{2}(7 + 28)(8)$
$= \frac{1}{2}(35)(8)$
$= 140 \text{ sq in}$

Circumference and area of a circle

Circumference (C) is the distance around the circle. The diameter (d) is the line segment that contains the center and has its end points on the circle. When the circumference of any circle is divided by its diameter, the result is always the same. That result is named after the Greek letter π (pi). The commonly used values for π are

$$\pi \approx 3.14 \text{ or } \pi \approx \frac{22}{7}$$

Use either value in your calculations. The formula for circumference is

$$C = \pi d \text{ or } C = 2\pi r$$

in which r = radius, a line segment from the center of the circle to one side, which is half the length of the diameter.

Example 16: What is the circumference of the circle shown in Figure 9-10?

Figure 9-10 Circle with center *M*.

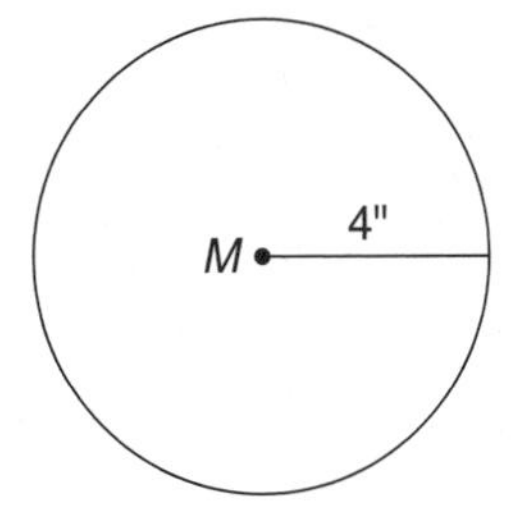

In the circle, $r = 4$, so $d = 8$.

$$C = \pi d$$
$$= \pi (8)$$
$$\approx 3.14(8) \text{ or } \frac{22}{7}(8)$$
$$25.12 \text{ in or } \approx 25.14 \text{ in}$$

The area (A) of a circle can be determined by

$$A = \pi r^2$$

Example 17: What is the area of the circle shown in Figure 9-11?

Figure 9-11 Circle with center *M*.

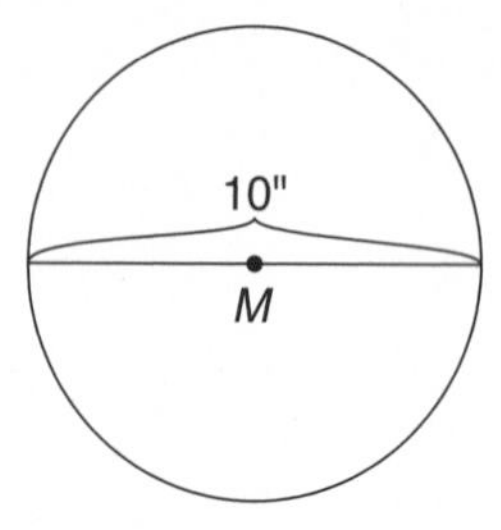

In the circle, $d = 10$, so $r = 5$.

$$A = \pi r^2$$
$$= \pi(5^2)$$
$$\approx 3.14(25) \text{ or } \frac{22}{7}(25)$$
$$78.5 \text{ sq in or} \approx 78.6 \text{ sq in}$$

Example 18: From the given radius or diameter, find the area and circumference (leave in terms of π) of the circles in Figure 9-12.

Figure 9-12 Circles with dimensions.

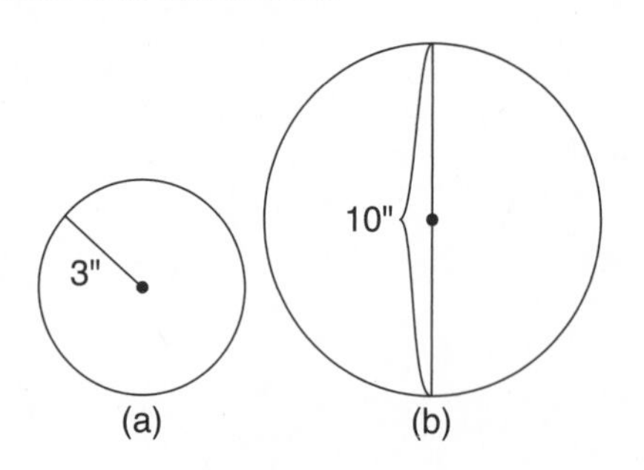

(a) $A = \pi r^2 = \pi(3^2) = 9\pi$ sq in; $C = 2\pi r = 2\pi(3) = 6\pi$ in or $C = \pi d = \pi(6) = 6\pi$ in

(b) $A = \pi r^2 = \pi(5^2) = 25\pi$ sq in; $C = 2\pi r = 2\pi(5) = 10\pi$ in or $C = \pi d = 10\pi$ in

Chapter Check-Out

1. How many feet are in one mile?
2. How many cups are in a pint?
3. True or false: *Hecto* means hundred in the metric system.
4. How many inches are in 4 yards?
5. If a kilometer equals 1,000 meters and 1 dekameter equals 10 meters, how many dekameters are in 5 kilometers?
6. Which of the following is more precise, 6.44 mm or 6.5 mm?
7. Find the number of units used and the significant digits for 35.6 m (unit of measure, 0.1 m).
8. What are the perimeter and area of the triangle?

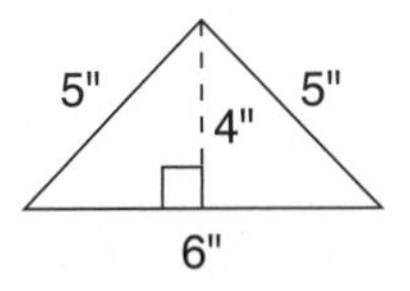

9. What are the perimeter and area of the parallelogram?

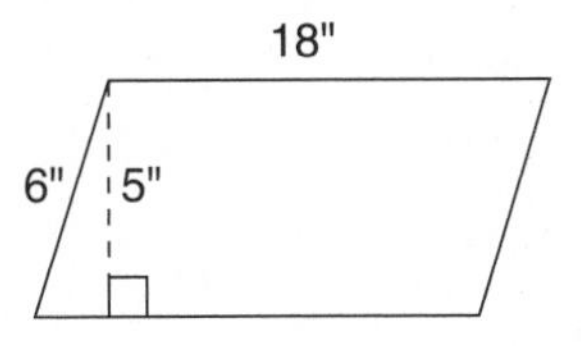

10. What are the circumference and area of the circle? Leave answers in π units.

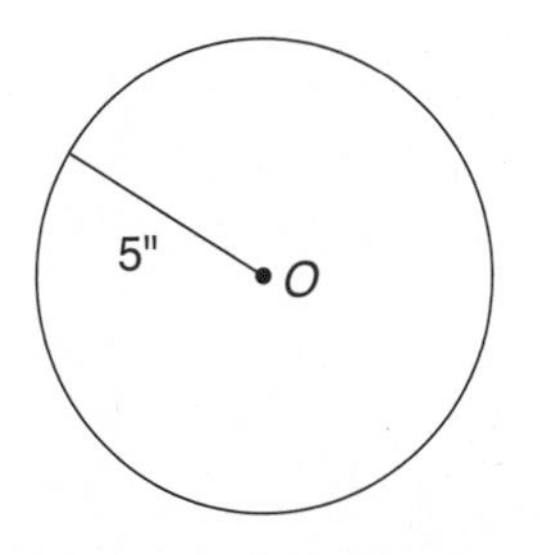

Answers: 1. 5,280 ft **2.** 2 cups **3.** True **4.** 144 inches **5.** 500 dekameters **6.** 6.44 mm **7.** 356 units, significant digits are 3, 5, 6 **8.** $P = 16$ in, $A = 12$ sq in **9.** $P = 48$ in, $A = 90$ sq in **10.** $C = 10\pi$ in, $A = 25\pi$ sq in

Chapter 10
GRAPHS

Chapter Check-In

- ❑ Bar graphs
- ❑ Line graphs
- ❑ Circle graphs (pie charts)
- ❑ Coordinate graphs

Information may be displayed in many ways. The three basic types of graphs you should know are bar graphs, line graphs, and circle graphs (or pie charts).

When answering questions related to a graph, you should

- Examine the entire graph—notice labels and headings.
- Focus on the information given.
- Look for major changes—high points, low points, trends.
- Do not memorize the graph; refer to it.
- Pay special attention to which part of the graph the question is referring to.
- Reread the headings and labels if you don't understand.

Bar Graphs

Bar graphs convert the information in a chart into separate bars or columns. Some graphs list numbers along one edge and places, dates, people, or things (individual categories) along another edge. Always try to determine the *relationship* between the columns in a graph or chart.

Example 1: The bar graph shown in Figure 10-1 indicates that City W has approximately how many more billboards than does City Y?

Figure 10-1 Horizontal bar graph of billboards in each city.

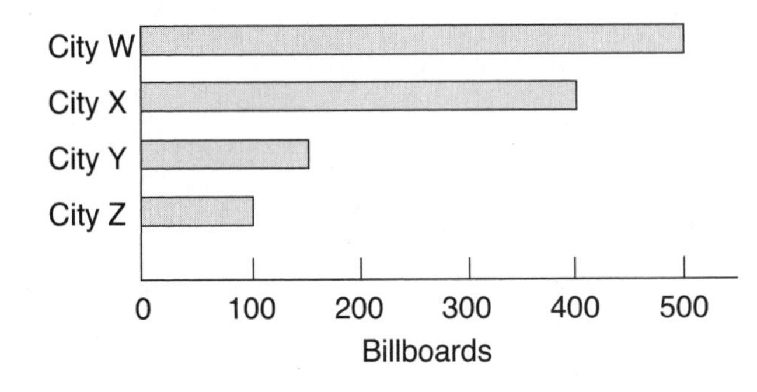

Notice that the graph shows the "Number of Billboards in Each City," with the numbers given along the bottom of the graph in increments of 100. The names are listed along the left side. City W has approximately 500 billboards. The bar graph for City Y stops about halfway between 100 and 200. Now, consider that halfway between 100 and 200 would be 150. So City W (500) has approximately 350 more billboards than does City Y (150).

$$500 - 150 = 350$$

Example 2: Based on the bar graph shown in Figure 10-2.

(a) The number of books sold by Mystery Mystery 1990–1992 exceeded the number of those sold by All Sports by approximately how many?

(b) From 1991 to 1992, the percent increase in number of books sold for All Sports exceeded the percent increase of Mystery Mystery by approximately how much?

(c) What caused the 1992 decline in Reference Unlimited's number of books sold?

Figure 10-2 Vertical bar graph showing the number of books sold by three publishers.

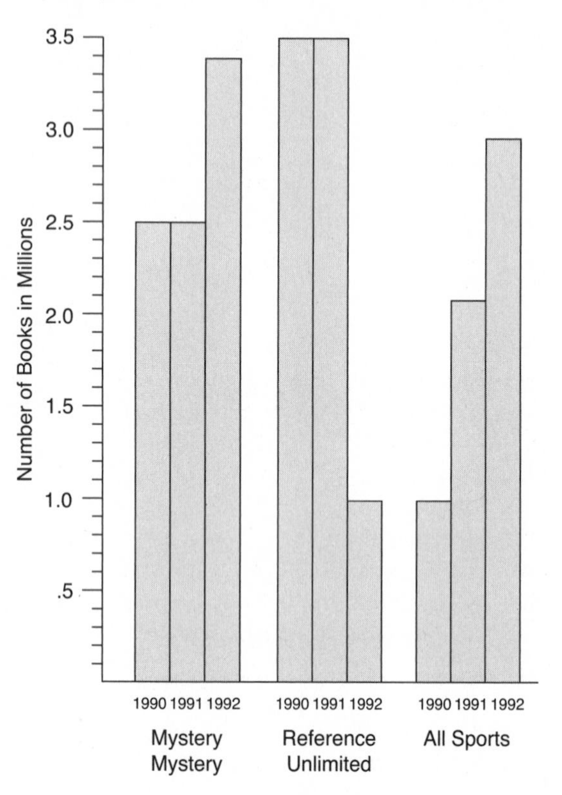

The graph in Figure 10-2 contains multiple bars representing each publisher. Each single bar stands for the number of books sold in a single year. You may be tempted to write out the numbers as you do your arithmetic (3.5 million = 3,500,000). Writing out the numbers is unnecessary, as it often is on graphs that use large numbers. Since *all* measurements are in millions, adding zeros does not add precision to the numbers.

(a) Referring to the Mystery Mystery bars, you can see that the number of books sold per year, in millions, is as follows:

1990 = 2.5

1991 = 2.5

1992 = 3.4

Use a piece of paper as a straightedge to determine this last number. Totaling the number of books sold for all three years gives 8.4 million.

Referring to the All Sports bars, you can see that number of books sold per year in millions is as follows:

1990 = 1

1991 = 2.1

1992 = 3

Once again, use a piece of paper as a straightedge, but don't designate numbers beyond the nearest tenth, since the graph numbers prescribe no greater accuracy than this. Totaling the number of books sold for all three years gives 6.1 million.

So, the number of books sold by Mystery Mystery exceeds the number of books sold by All Sports by 2.3 million.

(b) Graph and chart questions may ask you to calculate percent increase or percent decrease. As you learned in Chapter 5, the formula for figuring either of these is the same.

$$\frac{\text{change}}{\text{starting point}} \times 100\% = \text{percent change}$$

In this case, the percent increase in number of books sold by Mystery Mystery can be calculated first.

Number of books sold in 1991 (in millions) = 2.5

Number of books sold in 1992 (in millions) = 3.4

Change (in millions) = 0.9

The 1991 amount is the "starting point," so

$$\frac{\text{change}}{\text{starting point}} \times 100\% = \frac{0.9}{2.5} \times 100\% = 36\%$$

The percent increase in number of books sold by All Sports can be calculated as follows.

Number of books sold in 1991 (in millions) = 2.1

Number of books sold in 1992 (in millions) = 3

Change (in millions) = 0.9

$$\frac{\text{change}}{\text{starting point}} \times 100\% = \frac{0.9}{2.1} \times 100\% \approx 43\%$$

So the percent increase of All Sports exceeds that of Mystery Mystery by 7%:

$$43\% - 36\% = 7\%$$

(c) This question cannot be answered based on the information in the graph. Never assume information not given. In this case, the multiple factors that could cause a decline in number of books sold are not represented by the graph.

Line Graphs

Line graphs convert data into points on a grid. These points are then connected to show a relationship among the items, dates, times, and so on. Notice the slopes of the lines connecting the points. These lines show increases and decreases. The sharper the slope *upward,* the greater the *increase.* The sharper the slope *downward,* the greater the *decrease.* Line graphs can show trends, or changes, in data over a period of time.

Example 3: Based on the line graph shown in Figure 10-3,

(a) In what year was the property value of Moose Lake Resort about $500,000?

(b) In which ten-year period was there the greatest decrease in the property value of Moose Lake Resort?

Figure 10-3 Line graph showing the property value history of Moose Lake Resort.

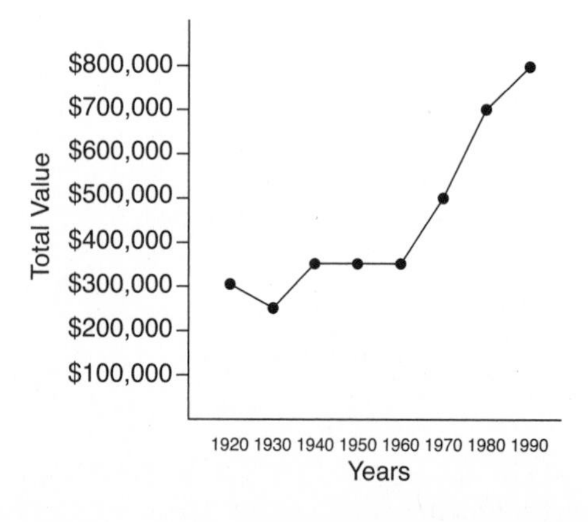

(a) The information along the left side of the graph shows the property value of Moose Lake Resort in increments of $100,000. The bottom of the graph shows the years from 1920 to 1990. Notice that in 1970 the property value was about $500,000. Using the edge of a sheet of paper as a straightedge helps you see that the dot in the 1970 column lines up with $500,000 on the left.

(b) Because the slope of the line goes *down* from 1920 to 1930, there must have been a decrease in property value. If you read the actual numbers, you notice a decrease from $300,000 to about $250,000.

Example 4: According to the line graph shown in Figure 10-4, the tomato plant grew the most between which two weeks?

Figure 10-4 Line graph of a tomato plant's growth.

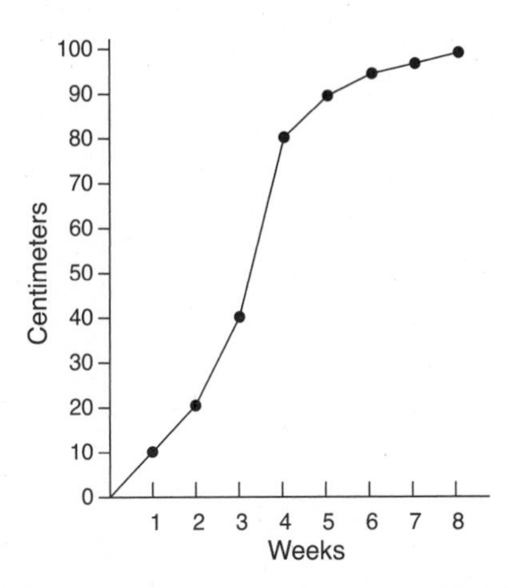

The numbers at the bottom of the graph give the weeks of growth of the plant. The numbers on the left give the height of the plant in centimeters. The sharpest upward slope occurs between week 3 and week 4, when the plant grew from 40 centimeters to 80 centimeters, a total of 40 centimeters growth.

Circle Graphs or Pie Charts

A circle graph, or pie chart, shows the relationship between the whole circle (100%) and the various slices that represent portions of that 100%. The larger the slice, the higher the percentage.

Example 5: Based on the circle graph shown in Figure 10-5,

(a) If Smithville Community Theater has \$1,000 to spend this month, how much will be spent on set construction?

(b) What is the ratio of the amount of money spent on advertising to the amount of money spent on set construction?

Figure 10-5 Circle graph.

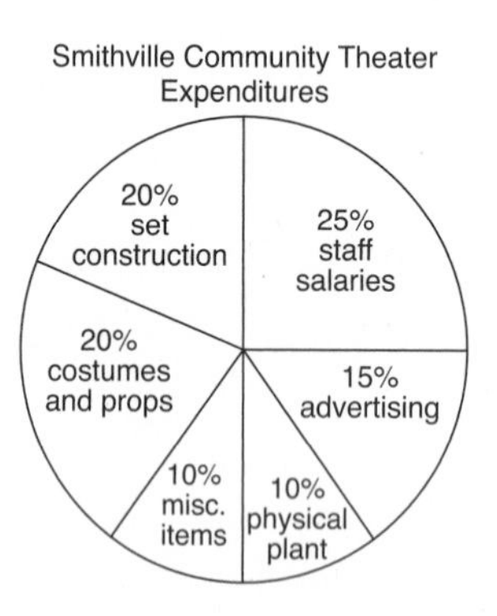

(a) The theater spends 20% of its money on set construction. 20% of \$1,000 is \$200, so \$200 will be spent on set construction.

(b) To answer this question, you must use the information in the graph to make a ratio.

$$\frac{\text{advertising}}{\text{set construction}} = \frac{15\% \text{ of } 1000}{20\% \text{ of } 1000} = \frac{150}{200} = \frac{3}{4}$$

Notice that $\frac{15\%}{20\%}$ simplifies to $\frac{3}{4}$.

Example 6: Based on the circle graph shown in Figure 10-6,

(a) If the Bell Canyon PTA spends the same percentage on dances every year, how much will they spend on dances in a year in which their total amount spent is \$15,000?

(b) The amount of money spent on field trips in 1995 was approximately what percent of the total amount spent?

Figure 10-6 Circle graph of 1995 budget of the Bell Canyon PTA.

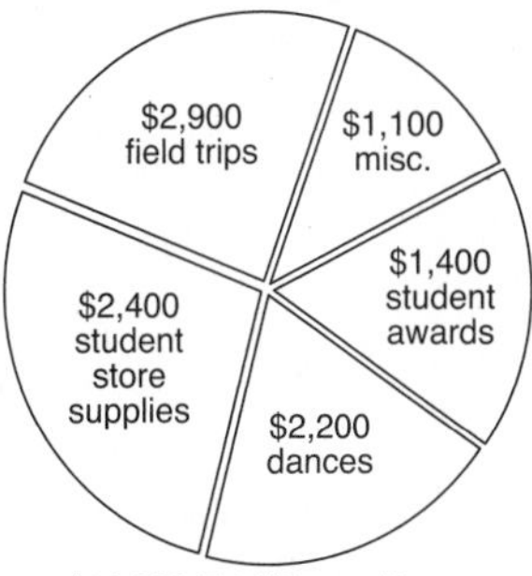

(a) To answer this question, you must find a percent and then apply this percent to a new total. In 1995, the PTA spent $2,200 on dances. This can be calculated to be 22% of the total spent in 1995 by the following method.

$$\frac{2{,}200}{10{,}000} = \frac{22}{100} = 22\%$$

Now, multiplying 22% times the *new* total amount spent of $15,000 gives the right answer.

$$22\% = 0.22$$

$$0.22 \times 15{,}000 = 3{,}300 \text{ or } \$3{,}300$$

You could use another common-sense method. If $2,200 out of $10,000 is spent for dances, $1,100 out of every $5,000 is spent for dances. Since $15,000 is 3 × $5,000, 3 × $1,100 would be $3,300.

(b) By carefully reading the information in the graph, you find that $2,900 was spent on field trips. The information describing the graph explains that the total expenditures were $10,000. The approximate percentage would be worked out as follows.

$$\frac{2{,}900}{10{,}000} = \frac{29}{100} = 29\%$$

Coordinate Graphs

Each point on a number line is assigned a number. In the same way, each point in a plane is assigned a pair of numbers. These numbers represent the placement of the point relative to two intersecting lines. In coordinate graphs (Figure 10-7), two perpendicular number lines are used and are called coordinate axes. One axis is horizontal and is called the x-axis. The other is vertical and is called the y-axis. The point of intersection of the two number lines is called the origin and is represented by the coordinates (0,0).

Figure 10-7 Coordinate graph.

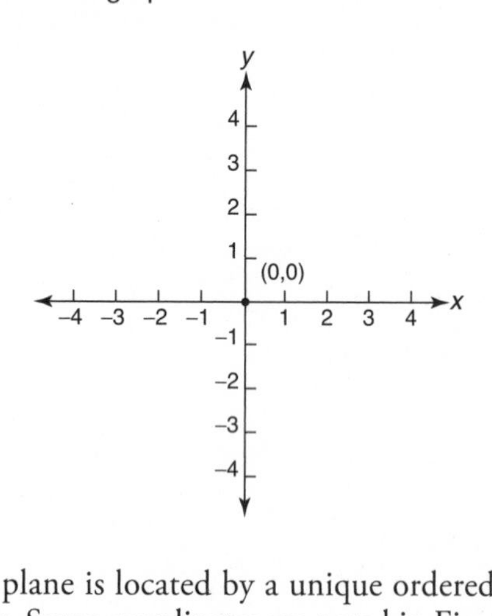

Each point on a plane is located by a unique ordered pair of numbers called coordinates. Some coordinates are noted in Figure 10-8.

Figure 10-8 Coordinate graph showing some coordinates.

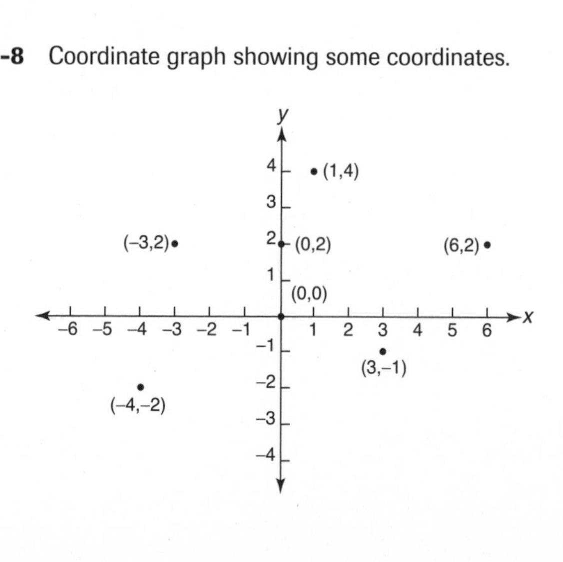

Notice that on the *x*-axis, numbers to the right of 0 are positive and to the left of 0 are negative. On the *y*-axis, numbers above 0 are positive and below 0 are negative. Also note that the first number in the ordered pair is called the *x*-coordinate, or abscissa, while the second number is the *y*-coordinate, or ordinate. The *x*-coordinate shows the right or left direction, and the *y*-coordinate shows the up or down direction.

The coordinate graph is divided into four quarters called quadrants. These quadrants are labeled in Figure 10-9.

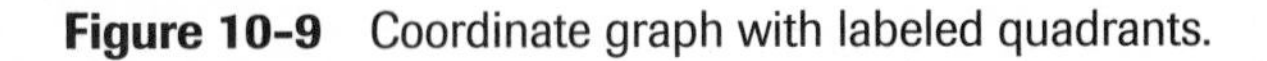

Figure 10-9 Coordinate graph with labeled quadrants.

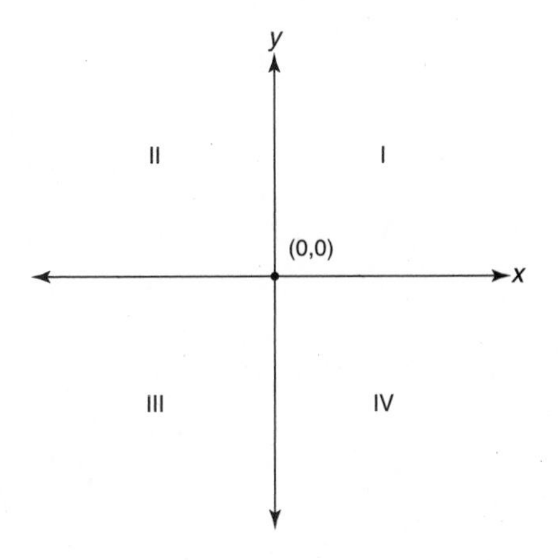

Notice that

- In quadrant I, *x* is always positive and *y* is always positive.
- In quadrant II, *x* is always negative and *y* is always positive.
- In quadrant III, *x* and y are both always negative.
- In quadrant IV, *x* is always positive and *y* is always negative.

Example 7: Identify the points (*A, B, C, D, E,* and *F*) on the coordinate graph shown in Figure 10-10.

Figure 10-10 Coordinate graph.

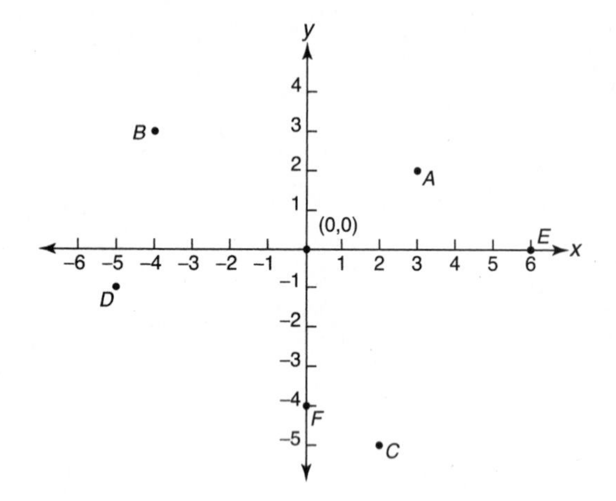

A (3, 2)
B (–4, 3)
C (2, –5)
D (–5, –1)
E (6, 0)
F (0, –4)

Chapter Check-Out

1. The bar graph indicates that Team A won how many more games than Team B?

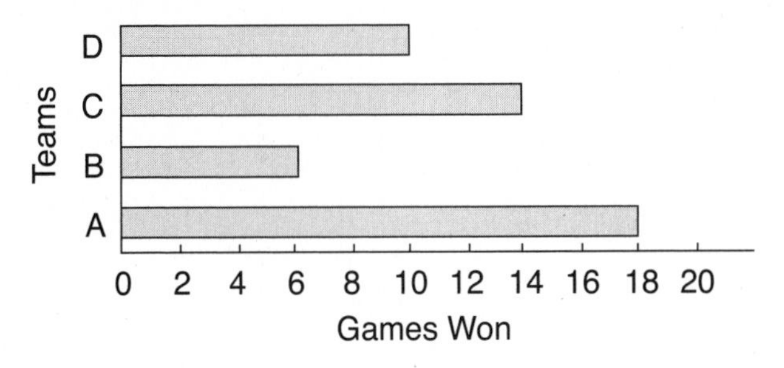

2. According to the line graph, Timmy grew the most between which two years?

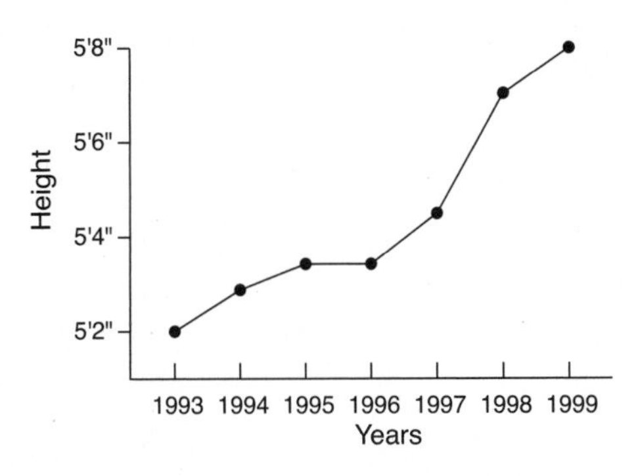

3. Based on the circle graph, how much time does Timmy spend doing homework?

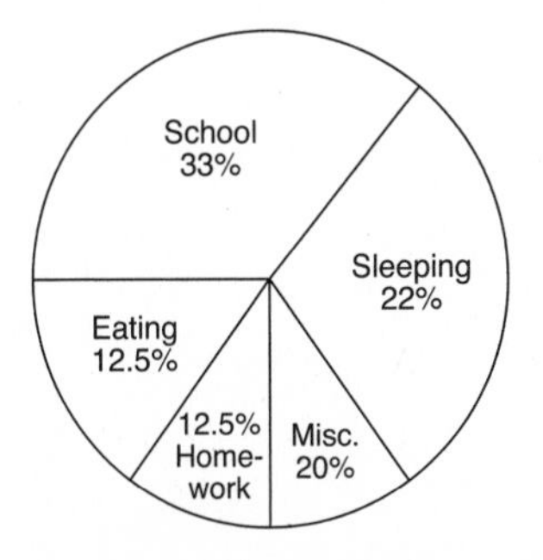

How Timmy Spends His 24-Hour Day

4. Identify the coordinates of the points A and B on the coordinate graph.

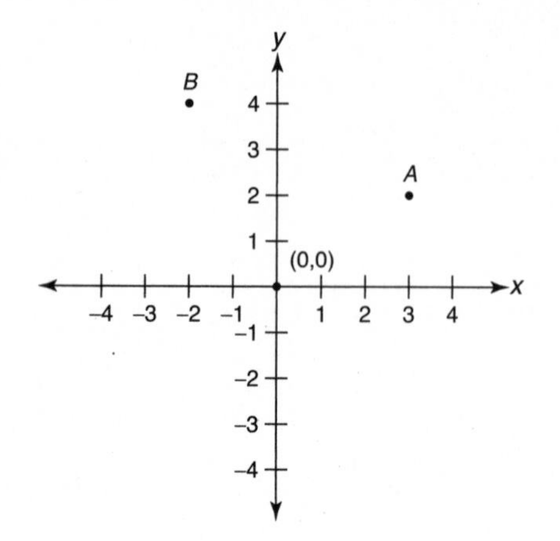

Answers: 1. 12 **2.** 1997 and 1998 **3.** 3 hours **4.** A (3,2) B (–2,4)

Chapter 11

PROBABILITY AND STATISTICS

Chapter Check-In

- ❑ Probability
- ❑ Combinations
- ❑ Permutations
- ❑ Mean
- ❑ Median
- ❑ Mode

Understanding the basics of probability and statistics will come in handy when you prepare for a number of different standardized math exams. However, before you can work with these concepts, you need to know some basic definitions.

Probability

Probability is the numerical measure of the chance of an outcome or event occurring. When all outcomes are equally likely to occur, the probability of the occurrence of a given outcome can be found by using the following formula:

$$\text{probability} = \frac{\text{number of favorable outcomes}}{\text{number of possible outcomes}}$$

Example 1: Using the spinner shown in Figure 11-1, what is the probability of spinning a 6 in one spin?

Figure 11-1 Spinner with equally divided sections.

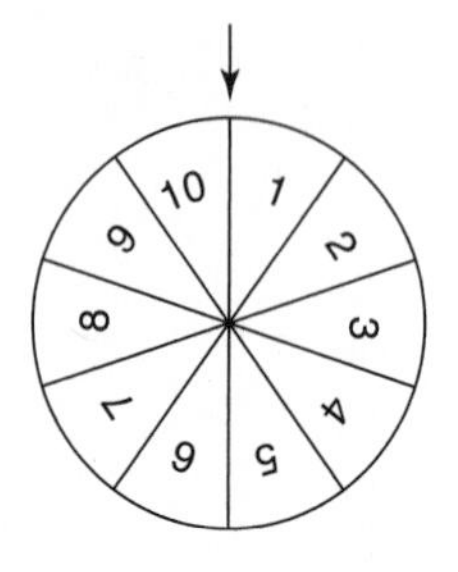

Since there is only *one* 6 on the spinner out of *ten* numbers and all the numbers are equally spaced, the probability is $\frac{1}{10}$.

Example 2: Again using the spinner shown in Figure 11-1, what is the probability of spinning either a 3 or a 5 in one spin?

Since there are *two favorable outcomes* out of *ten possible outcomes*, the probability is $\frac{2}{10}$ or $\frac{1}{5}$.

When two events are *independent* of each other, you need to multiply to find the favorable and/or possible outcomes.

Example 3: What is the probability that both of the spinners shown in Figure 11-2 will stop on a 3 on the first spin?

Figure 11-2 Spinners with equally divided sections.

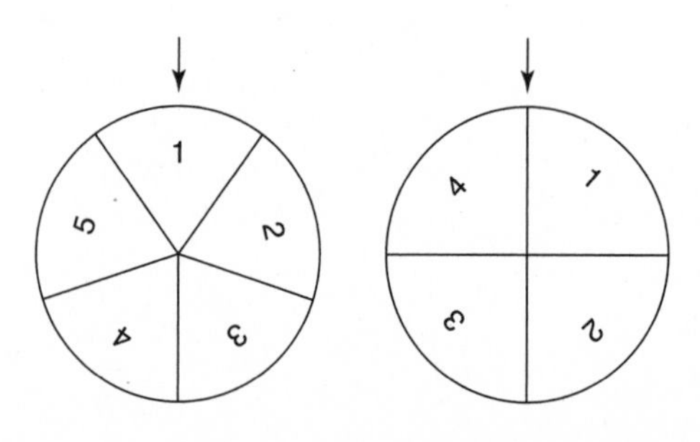

Since the probability that the first spinner will stop on the number 3 is $\frac{1}{5}$ and the probability that the second spinner will stop on the number 3 is $\frac{1}{4}$, and because each event is independent of the other, simply multiply.

$$\frac{1}{5} \times \frac{1}{4} = \frac{1}{20}$$

Example 4: What is the probability that on two consecutive rolls of a die the numbers will be 2 and then 3? (A die has 6 sides numbered 1–6.)

Since the probability of getting a 2 on the first roll is $\frac{1}{6}$ and the probability of getting a 3 on the second roll is $\frac{1}{6}$, and since the rolls are independent of each other, simply multiply.

$$\frac{1}{6} \times \frac{1}{6} = \frac{1}{36}$$

Example 5: What is the probability of tossing heads three consecutive times with a two-sided fair coin?

Because each toss is independent and the probability is $\frac{1}{2}$ for each toss, the probability is

$$\frac{1}{2} \times \frac{1}{2} \times \frac{1}{2} = \frac{1}{8}$$

Example 6: What is the probability of rolling two dice in one toss so that they total 5?

Since there are six possible outcomes on each die, the total possible outcomes for two dice is

$$6 \times 6 = 36$$

The favorable outcomes are (1 + 4), (4 +1), (2 + 3), and (3 + 2). These are all the ways of tossing a total of 5 on two dice. Thus, there are four favorable outcomes, which give the probability of throwing a total of five as

$$\frac{4}{36} = \frac{1}{9}$$

Example 7: Three green marbles, two blue marbles, and five yellow marbles are placed in a jar. What is the probability of selecting at random a green marble on the first draw?

Since there are ten marbles (total possible outcomes) and three green marbles (favorable outcomes), the probability is $\frac{3}{10}$.

Example 8: In a regular deck of 52 cards, what is the probability of drawing a heart on the first draw? (There are 13 hearts in a deck.)

Since there are 13 favorable outcomes out of 52 possible outcomes, the probability is $\frac{13}{52}$ or $\frac{1}{4}$.

Arrangements

If there are a *number of successive choices* to make and the choices are *independent of each other* (order makes no difference), the total number of possible choices is the product of each of the choices at each stage.

Example 9: How many possible combinations of shirts and ties are there if there are five different color shirts and three different color ties?

To find the total number of possible combinations, simply multiply the number of shirts times the number of ties.

$$5 \times 3 = 15$$

Example 10: A combination lock has three settings, each of which contains numbers from 0 to 9. How many different possible combinations exist on the lock?

Note that each setting is independent of the others; thus, because each has ten possible settings,

$$10 \times 10 \times 10 = 1{,}000$$

There are 1,000 possible combinations.

Permutations

If there are a *number of successive choices* to make and the choices are *affected by the previous choice or choices* (dependent upon order), then **permutations** are involved.

Example 11: How many ways can you arrange the letters *S, T, O, P* in a row?

number of choices for the first letter		number of choices for the second letter		number of choices for the third letter		number of choices for the fourth letter
4	×	3	×	2	×	1

$$4! = 4\times3\times2\times1 = 24$$

The product $4 \times 3 \times 2 \times 1$ can be written 4! (read *4 factorial or factorial 4*). Thus, there are 24 different ways to arrange four different letters.

Example 12: How many different ways are there to arrange three jars in a row on a shelf?

Because the order of the items is affected by the previous choice(s), the number of different ways equals 3! or

$$3\times2\times1 = 6$$

There are six different ways to arrange the three jars.

Example 13: If, from among five people, three executives are to be selected, how many possible combinations of executives are there?

This is a more difficult type of arrangement involving permutations. Notice here that the order of selection makes no difference. The symbol used to denote this situation is

C(*n, r*), which is read *the number of combinations of n things taken r at a time.* The formula used is

$$C(n,r) = \frac{n!}{r!\left(n-r\right)!}$$

Because $n = 5$ and $r = 3$ (five people taken three at a time), then the solution is as follows:

$$\frac{5!}{3!(5-3)!}$$

Now solve.

$$\frac{5!}{3!(5-3)!} = \frac{5\cdot4\cdot3\cdot2\cdot1}{3\cdot2\cdot1\cdot(2)!} = \frac{5\cdot\overset{2}{\cancel{4}}\cdot\overset{1}{\cancel{3}}\cdot\overset{1}{\cancel{2}}\cdot1}{\underset{1}{\cancel{3}}\cdot\underset{1}{\cancel{2}}\cdot1\cdot(\underset{1}{\cancel{2}}\cdot1)} = 10$$

If the problem involves very few possibilities, you may want to actually list the possible combinations.

Example 14: A coach is selecting a starting lineup for her basketball team. She must select from among nine players to get her starting lineup of five. How many possible starting lineups could she have?

Because $n = 9$ and $r = 5$ (nine players taken five at a time), the solution is as follows.

$$\frac{9!}{5!(9-5)!}$$

$$\frac{9\cdot8\cdot7\cdot6\cdot5\cdot4\cdot3\cdot2\cdot1}{5\cdot4\cdot3\cdot2\cdot1\cdot(4)!}=\frac{9\cdot\overset{2}{\cancel{8}}\cdot7\cdot\overset{1}{\cancel{6}}\cdot\overset{1}{\cancel{5}}\cdot\overset{1}{\cancel{4}}\cdot\overset{1}{\cancel{3}}\cdot\overset{1}{\cancel{2}}\cdot1}{\underset{1}{\cancel{5}}\cdot\underset{1}{\cancel{4}}\cdot\underset{1}{\cancel{3}}\cdot2\cdot1\cdot(\underset{1}{\cancel{4}}\cdot\underset{1}{\cancel{3}}\cdot\underset{1}{\cancel{2}}\cdot1)}=126$$

Example 15: How many possible combinations of *a*, *b*, *c*, and *d*, taken two at a time, are there?

Since $n = 4$ and $r = 2$ (four letters taken two at a time), the solution is as follows:

$$\frac{4!}{2!(4-2)!}=\frac{4!}{2!(2)!}=\frac{\overset{2}{\cancel{4}}\cdot3\cdot\overset{1}{\cancel{2}}\cdot1}{\underset{1}{\cancel{2}}\cdot1\cdot(\underset{1}{\cancel{2}}\cdot1)}=6$$

You may simply have listed the possible combinations as *ab, ac, ad, bc, bd,* and *cd.*

Statistics

The study of numerical data and their distribution is called statistics.

Any measure indicating a center of a distribution is called a measure of central tendency. The three basic measures of central tendency are

- **Mean (or arithmetic mean):** Usually called the average.
- **Median:** The middle number in a set of numbers arranged in ascending or descending order.
- **Mode:** The set, class, or classes that show up most often.

Mean

The mean (arithmetic mean) is the most frequently used measure of central tendency. It is generally reliable, easy to use, and more stable than the median. To determine the arithmetic mean, simply total the items and then divide by the number of items.

Example 16: What is the arithmetic mean of 0, 12, 18, 20, 31, and 45?

Total the items.

$$0 + 12 + 18 + 20 + 31 + 45 = 126$$

Divide by the number of items.

$$126 \div 6 = 21$$

The arithmetic mean is 21.

Example 17: What is the arithmetic mean of 25, 27, 27, and 27?

$$25 + 27 + 27 + 27 = 106$$

$$106 \div 4 = 26\frac{1}{2}$$

The arithmetic mean is $26\frac{1}{2}$.

Example 18: What is the arithmetic mean of 20 and –10?

$$20 + (-10) = +10$$

$$10 \div 2 = 5$$

The arithmetic mean is 5.

When one or a number of items is used several times, those items have more "weight." This establishment of relative importance, or *weighting*, is used to compute the **weighted mean.**

Example 19: What is the mean of three tests averaging 70% plus seven tests averaging 85%?

In effect, you have ten exams, three of which score 70% and seven of which score 85%. Rather than adding all ten scores, to determine the above "weighted mean," simply multiply 3 times 70% to find the total of those items (210). Then multiply 7 times 85% to find their total (595). Now add the two totals (805) and divide by the number of items overall (10). The weighted mean is thus 80.5%.

Example 20: For the first nine months of the year, the average monthly rainfall was 2 inches. For the last three months of that year, rainfall averaged 4 inches per month. What was the mean monthly rainfall for the entire year?

$$\begin{array}{l} 9 \times 2'' = 18'' \\ \underline{3 \times 4'' = 12''} \\ \text{total} = 30'' \text{ divided by 12 months in all } = 2.5'' \text{ monthly mean} \end{array}$$

Example 21: Six students averaged 90% on a class test. Four other students averaged 70% on the test. What was the mean score of all ten students?

$$\begin{array}{l} 6 \times 90\% = 540\% \\ \underline{4 \times 70\% = 280\%} \\ \text{total} = 820\% \text{ divided by 10 students} = 82\% \end{array}$$

Median

The median of a set of numbers arranged in ascending or descending order is the middle number if there is an odd number of items in the set. If there is an even number of items in the set, their median is the arithmetic mean of the middle two numbers. The median is easy to calculate and is not influenced by extreme measurements.

Example 22: Find the median of 3, 4, 6, 9, 21, 24, 56.

$$3, 4, 6, \underline{9}, 21, 24, 56$$

The median is 9.

Example 23: Find the median of 4, 5, 6, 10.

$$4, \underbrace{5, 6}, 10$$

$$\frac{5+6}{2} = 5\frac{1}{2}$$

The median is $5\frac{1}{2}$.

Mode

The set, class, or classes that appear most, or whose frequency is the greatest, is the mode or modal class. (Mode is not greatly influenced by extreme cases but is probably the least important or least used of the three measures of central tendency.)

Example 24: Find the mode of 3, 4, 8, 9, 9, 2, 6, 11.

The mode is 9 because it appears more often than any other number.

Chapter Check-Out

1. Using the equally spaced spinner, what is the probability of spinning either a 5 or a 6 in one spin?

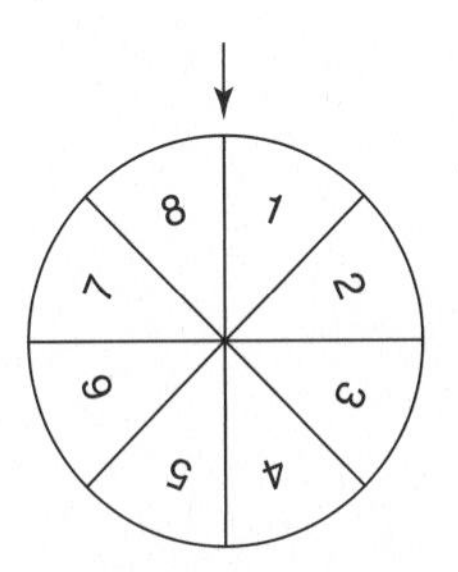

2. What is the probability that on two consecutive rolls of a die the numbers will be 5 and then 6?
3. What is the probability of tossing tails four consecutive times with a two-sided fair coin?
4. What is the probability of rolling two dice in one toss so that they total 7?
5. Five gold coins, two silver coins, and seven copper coins, all of the same size, are placed in a bag. What is the probability of selecting at random a gold coin on the first draw?
6. How many possible arrangements of shirts and pants are there if there are eight different color shirts and three different color pants?
7. How many ways can you arrange the letters *S, A, L, E* in a row?
8. If four researchers are to be selected for a project from a group of six researchers, how many possible combinations of researchers for the project are there?

9. What is the arithmetic mean, mode, and median of the following numbers: 3, 5, 4, 3, 4, 4, 6, 7, 2?

10. What is the mean weight of seven men if five men weigh 160 pounds and two men 180 pounds?

Answers: 1. $\frac{2}{8}$ or $\frac{1}{4}$ **2.** $\frac{1}{36}$ **3.** $\frac{1}{16}$ **4.** $\frac{6}{36}$ or $\frac{1}{6}$ **5.** $\frac{5}{14}$ **6.** 24 7. 24 **8.** 15 **9.** mean = $4\frac{2}{9}$, mode = 4, median = 4 **10.** $165\frac{5}{7}$ pounds

Chapter 12
NUMBER SERIES

Chapter Check-In

- ❑ Arithmetic progressions
- ❑ Geometric progressions

A **number series** is a mathematical progression of numbers with some pattern. And although some sequences are simply patterns (2, 2, 4, 5, 2, 2, 6, 7, 2, 2, 8, 9, . . .), the two common types of mathematical sequences are arithmetic progressions and geometric progressions.

Arithmetic Progressions

An arithmetic progression is a progression in which there is a common difference between terms. Subtract any term from the next, and you get the same value.

Example 1: What is the next number in the progression 4, 7, 10, 13, . . .?

Since the common difference is 3, the next number is 16.

Example 2: What is the 40th term in the progression 5, 10, 15, 20, . . .?

The common difference is 10 – 5 = 5. The 40th term is 36 terms beyond the fourth term, 20.

$$5 \times 36 = 180$$

$$\underbrace{20 + 180}_{40\text{th term}}$$

$$5, 10, 15, 20, \ldots, 200$$

So, the 40th term is 200.

Geometric Progressions

A geometric progression is a progression in which there is a constant ratio between terms. The value found by dividing any term by the term before it gives you the constant ratio.

Example 3: What is the next number in the progression 1, 2, 4, 8, 16, . . .?

The ratio is= $\frac{2}{1}$ because $\frac{16}{8}=\frac{8}{4}=\frac{4}{2}=\frac{2}{1}$. You could also have said that you're simply doubling each number to get the next number. So doubling the last number, 16, gives 32.

Example 4: What is the next number in the progression 3, 9, 27, 81, . . .?

The ratio is $\frac{3}{1}$ because $\frac{81}{27}=\frac{27}{9}=\frac{9}{3}=\frac{3}{1}$. You could also have said you are multiplying each number by 3 to get the next number. So multiply the last number, 81, by 3, which gives 243.

Chapter Check-Out

1. What is the next number in the progression
   5, 9, 13, 17, . . .?
2. What is the 30th term in the progression
   4, 8, 12, 16, . . .?
3. What is the next number in the progression
   512, 128, 32, 8, . . .?
4. What is the next number in the progression
   5, 25, 125, 625, . . .?

Answers: 1. 21 **2.** 120 **3.** 2 **4.** 3,125

Chapter 13

VARIABLES, ALGEBRAIC EXPRESSIONS, AND SIMPLE EQUATIONS

Chapter Check-In

- ❑ Translating algebraic expressions
- ❑ Evaluating expressions
- ❑ Solving simple equations
- ❑ Checking your answers

As you get ready to move from pre-algebra to algebra, you should feel comfortable working with variables and algebraic expressions. You should also be able to solve some simple equations.

Variables and Algebraic Expressions

Before you start solving equations, you should have a basic understanding of variables, as well as translating and evaluating algebraic expressions.

Variables

A variable is a letter used to stand for a number. The letters x, y, z, a, b, c, m, and n are probably the most commonly used variables. The letters e and i have special values in algebra and are usually not used as variables. The letter o is usually not used because it can be mistaken for 0 (zero).

Algebraic expressions

Variables are used to change verbal expressions into algebraic expressions, that is, expressions that are composed of letters that stand for numbers. Key words that can help you translate words into letters and numbers include:

- *For addition:* sum, more than, greater than, increase
- *For subtraction:* minus, less than, smaller than, decrease
- *For multiplication:* times, product, multiplied by, of
- *For division:* halve, divided by, ratio.

Also see "Key Words," in Chapter 14.

Example 1: Give the algebraic expression for each of the following.

(a) the sum of a number and 5

(b) the number minus 4

(c) six times a number

(d) x divided by 7

(e) three more than the product of 2 and x

(a) the sum of a number and 5: $x + 5$ or $5 + x$

(b) the number minus 4: $x - 4$

(c) six times a number: $6x$

(d) x divided by 7: $x / 7$ or $\frac{x}{7}$

(e) three more than the product of 2 and x: $2x + 3$

Evaluating expressions

To evaluate an expression, just replace the variables with grouping symbols, insert the values given for the variables, and do the arithmetic. Remember to follow the order of the operations: parentheses, exponents, multiplication/division, addition/subtraction.

Example 2: Evaluate each of the following.

(a) $x + 2y$ if $x = 2$ and $y = 5$

(b) $a + bc - 3$ if $a = 4$, $b = 5$, and $c = 6$

(c) $m^2 + 4n + 1$ if $m = 3$ and $n = 2$

(d) $\frac{b+c}{7} + \frac{c}{a}$ if $a = 2$, $b = 3$, and $c = 4$

(e) $-5xy + z$ if $x = 6$, $y = 7$, and $z = 1$

(a) $x + 2y = (2) + 2(5)$

$= 2 + 10$

$= 12$

(b) $a + bc - 3 = (4) + (5)(6) - 3$

$= 4 + 30 - 3$

$= 34 - 3$

$= 31$

(c) $m^2 + 4n + 1 = (3)^2 + 4(2) + 1$

$= 9 + 8 + 1$

$= 17 + 1$

$= 18$

(d) $\frac{b+c}{7} + \frac{c}{a} = \frac{(3)+(4)}{7} + \frac{(4)}{(2)}$

$= \frac{7}{7} + 2$

$= 1 + 2$

$= 3$

(e) $-5xy + z = -5(6)(7) + (1)$

$= -5(42) + 1$

$= -210 + 1$

$= -209$

Solving Simple Equations

When solving a simple equation, think of the equation as a balance, with the equals sign (=) being the fulcrum or center. Thus, if you do something to one side of the equation, you must do the same thing to the other side. Doing the *same thing to both sides* of the equation (say, adding 3 to each side) keeps the equation balanced.

Solving an equation is the process of getting what you're looking for, or *solving for,* on one side of the equals sign and everything else on the other side. You're really sorting information. If you're solving for x, you must get x on one side by itself.

Addition and subtraction equations

Some equations involve only addition and/or subtraction.

Example 3: Solve for x.

$$x + 8 = 12$$

To solve the equation $x + 8 = 12$, you must get x by itself on one side. Therefore, subtract 8 from both sides.

$$\begin{array}{rcl} x+8 &=& 12 \\ \underline{\quad -8} && \underline{-8} \\ x &=& 4 \end{array}$$

To check your answer, simply plug your answer into the equation:

$$\begin{aligned} x+8 &= 12 \\ (4)+8 &\stackrel{?}{=} 12 \\ 12 &= 12 \checkmark \end{aligned}$$

Example 4: Solve for y.

$$y - 9 = 25$$

To solve this equation, you must get y by itself on one side. Therefore, add 9 to both sides.

$$\begin{array}{rcl} y-9 &=& 25 \\ \underline{\quad +9} && \underline{+9} \\ y &=& 34 \end{array}$$

To check, simply replace y with 34:

$$y - 9 = 25$$
$$(34) - 9 \stackrel{?}{=} 25$$
$$25 = 25 \checkmark$$

Example 5: Solve for x.

$$x + 15 = 6$$

To solve, subtract 15 from both sides.

$$\begin{aligned} x + 15 &= 6 \\ -15 &= -15 \\ \hline x \quad &= -9 \end{aligned}$$

To check, simply replace x with –9 :

$$x + 15 = 6$$
$$(-9) + 15 \stackrel{?}{=} 6$$
$$6 = 6 \checkmark$$

Notice that in each case above, *opposite operations* are used; that is, if the equation has addition, you subtract from each side.

Multiplication and division equations

Some equations involve only multiplication or division. This is typically when the variable is already on one side of the equation, but there is either more than one of the variable, such as $2x$, or a fraction of the variable, such as

$$\frac{x}{3} \text{ or } \left(\frac{1}{2}\right)x$$

In the same manner as when you add or subtract, you can multiply or divide both sides of an equation by the same number, *as long as it is not zero,* and the equation will not change.

Example 6: Solve for x.

$$3x = 9$$

Divide each side of the equation by 3.

$$3x = 9$$

$$\frac{3x}{3} = \frac{9}{3}$$

$$\frac{\overset{1}{\cancel{3}}x}{\underset{1}{\cancel{3}}} = \frac{9}{3}$$

$$x = 3$$

To check, replace x with 3:

$$3x = 9$$

$$3(3) \overset{?}{=} 9$$

$$9 = 9 \checkmark$$

Example 7: Solve for y.

$$\frac{y}{5} = 7$$

To solve, multiply each side by 5.

$$\frac{y}{5} = 7$$

$$(5)\left(\frac{y}{5}\right) = (7)(5)$$

$$\left(\frac{\overset{1}{\cancel{5}}}{1}\right)\left(\frac{y}{\underset{1}{\cancel{5}}}\right) = 35$$

$$y = 35$$

To check, replace y with 35:

$$\frac{y}{5} = 7$$
$$\frac{35}{5} \stackrel{?}{=} 7$$
$$7 = 7 \checkmark$$

Example 8: Solve for x.

$$\frac{3}{4}x = 18$$

To solve, multiply each side by $\frac{4}{3}$.

$$\frac{3}{4}x = 18$$
$$\left(\frac{4}{3}\right)\left(\frac{3}{4}x\right) = (18)\left(\frac{4}{3}\right)$$
$$\left(\frac{\overset{1}{\cancel{4}}}{\underset{1}{\cancel{3}}}\right)\left(\frac{\overset{1}{\cancel{3}}}{\underset{1}{\cancel{4}}}x\right) = \left(\frac{\overset{6}{\cancel{18}}}{1}\right)\left(\frac{4}{\underset{1}{\cancel{3}}}\right)$$
$$x = 24$$

Or, without canceling,

$$\frac{3}{4}x = 18$$
$$\left(\frac{4}{3}\right)\left(\frac{3}{4}x\right) = (18)\left(\frac{4}{3}\right)$$
$$\frac{12}{12}x = \left(\frac{18}{1}\right)\left(\frac{4}{3}\right)$$

Notice that on the left you would normally not write $\frac{12}{12}$ because it would always cancel to $1x$, or x.

$$x = \frac{72}{3}$$
$$x = 24$$

Combinations of operations

Sometimes you have to use more than one step to solve the equation. In most cases, do the addition or subtraction step first. Then, after you've sorted the variables to one side and the numbers to the other, multiply or divide to get only one of the variables (that is, a variable with no number, or 1, in front of it: x, not $2x$).

Example 9: Solve for x.

$$2x + 4 = 10$$

Subtract 4 from both sides to get $2x$ by itself on one side.

$$\begin{array}{rcl} 2x + 4 & = & 10 \\ -4 & = & -4 \\ \hline 2x \quad & = & 6 \end{array}$$

Then divide both sides by 2 to get x.

$$2x = 6$$

$$\frac{2x}{2} = \frac{6}{2}$$

$$\frac{\overset{1}{\cancel{2}}x}{\underset{1}{\cancel{2}}} = \frac{6}{2}$$

$$x = 3$$

To check, substitute your answer into the original equation:

$$2x + 4 = 10$$

$$2(3) + 4 \overset{?}{=} 10$$

$$6 + 4 \overset{?}{=} 10$$

$$10 = 10 \checkmark$$

Example 10: Solve for *x*.

$$5x - 11 = 29$$

Add 11 to both sides.

$$\begin{aligned} 5x - 11 &= 29 \\ +11 &\quad +11 \\ \hline 5x &= 40 \end{aligned}$$

Divide each side by 5.

$$5x = 40$$

$$\frac{5x}{5} = \frac{40}{5}$$

$$\frac{\overset{1}{\cancel{5}}x}{\underset{1}{\cancel{5}}} = \frac{40}{5}$$

$$x = 8$$

To check, replace *x* with 8:

$$5x - 11 = 29$$

$$5(8) - 11 \overset{?}{=} 29$$

$$40 - 11 \overset{?}{=} 29$$

$$29 = 29 \checkmark$$

Example 11: Solve for *x*.

$$\frac{2}{3}x + 6 = 12$$

Subtract 6 from each side.

$$\begin{aligned} \frac{2}{3}x + 6 &= 12 \\ -6 &\quad -6 \\ \hline \frac{2}{3}x &= 6 \end{aligned}$$

Multiply each side by $\frac{3}{2}$.

$$\frac{2}{3}x = 6$$

$$\left(\frac{3}{2}\right)\left(\frac{2}{3}x\right) = 6\left(\frac{3}{2}\right)$$

$$\left(\frac{\overset{1}{\cancel{3}}}{\underset{1}{\cancel{2}}}\right)\left(\frac{\overset{1}{\cancel{2}}}{\underset{1}{\cancel{3}}}x\right) = \left(\frac{\overset{3}{\cancel{6}}}{1}\right)\left(\frac{3}{\underset{1}{\cancel{2}}}\right)$$

$$x = 9$$

To check, replace x with 9:

$$\frac{2}{3}x + 6 = 12$$

$$\frac{2}{3}(9) + 6 \stackrel{?}{=} 12$$

$$\left(\frac{2}{\underset{1}{\cancel{3}}}\right)\left(\frac{\overset{3}{\cancel{9}}}{1}\right) + 6 \stackrel{?}{=} 12$$

$$6 + 6 \stackrel{?}{=} 12$$

$$12 = 12\checkmark$$

Example 12: Solve for y.

$$\frac{2}{5}y - 8 = -18$$

Add 8 to both sides.

$$\begin{array}{rcl} \frac{2}{5}y - 8 & = & -18 \\ \underline{\quad +8} & & \underline{+8} \\ \frac{2}{5}y \quad & = & -10 \end{array}$$

Multiply each side by $\frac{5}{2}$.

$$\frac{2}{5}y = -10$$

$$\left(\frac{5}{2}\right)\left(\frac{2}{5}y\right) = -(10)\left(\frac{5}{2}\right)$$

$$\left(\frac{\cancel{5}^{1}}{\cancel{2}_{1}}\right)\left(\frac{\cancel{2}^{1}}{\cancel{5}_{1}}y\right) = \left(-\frac{\cancel{10}^{5}}{1}\right)\left(\frac{5}{\cancel{2}_{1}}\right)$$

$$y = -25$$

To check, replace y with -25:

$$\frac{2}{5}y - 8 = -18$$

$$\frac{2}{5}(-25) - 8 \stackrel{?}{=} -18$$

$$\left(\frac{2}{\cancel{5}_{1}}\right)\left(-\frac{\cancel{25}^{5}}{1}\right) - 8 \stackrel{?}{=} -18$$

$$-10 - 8 \stackrel{?}{=} -18$$

$$-18 = -18 \checkmark$$

Example 13: Solve for *x*.

$$3x + 2 = x + 4$$

Subtract 2 from both sides (which is the same as adding –2).

$$\begin{array}{rl} 3x + 2 & = x + 4 \\ -2 & \quad\;\; -2 \\ \hline 3x \quad & = x + 2 \end{array}$$

Subtract x from both sides.

$$\begin{array}{r} 3x = x+2 \\ \underline{-x \quad -x} \\ 2x = \quad 2 \end{array}$$

Note that $3x - x$ is the same as $3x - 1x$.

Divide both sides by 2.

$$2x = 2$$

$$\frac{2x}{2} = \frac{2}{2}$$

$$\frac{\overset{1}{\cancel{2}}x}{\underset{1}{\cancel{2}}} = \frac{2}{2}$$

$$x = 1$$

To check, replace x with 1:

$$3x + 2 = x + 4$$

$$3(1) + 2 \overset{?}{=} (1) + 4$$

$$3 + 2 = 1 + 4$$

$$5 = 5 \checkmark$$

Example 14: Solve for y.

$$5y + 3 = 2y + 9$$

Subtract 3 from both sides.

$$\begin{array}{r} 5y+3 = 2y+9 \\ \underline{-3 \quad\quad -3} \\ 5y \quad = 2y+6 \end{array}$$

Subtract $2y$ from both sides.

$$\begin{array}{r} 5y = 2y+6 \\ \underline{-2y \quad -2y} \\ 3y = \quad 6 \end{array}$$

Divide both sides by 3.

$$3y = 6$$

$$\frac{3y}{3} = \frac{6}{3}$$

$$\frac{\overset{1}{\cancel{3}}y}{\underset{1}{\cancel{3}}} = \frac{6}{3}$$

$$y = 2$$

To check, replace y with 2:

$$5y + 3 = 2y + 9$$

$$5(2) + 3 \stackrel{?}{=} 2(2) + 9$$

$$10 + 3 \stackrel{?}{=} 4 + 9$$

$$13 = 13 \checkmark$$

Sometimes you need to simplify each side (combine like terms) before actually starting the sorting process.

Example 15: Solve for x.

$$3x + 4 + 2 = 12 + 3$$

First, simplify each side.

$$3x + 4 + 2 = 12 + 3$$

$$3x + 6 = 15$$

Subtract 6 from both sides.

$$\begin{array}{rcr} 3x + 6 & = & 15 \\ -6 & & -6 \\ \hline 3x & = & 9 \end{array}$$

Divide both sides by 3.

$$3x = 9$$
$$\frac{3x}{3} = \frac{9}{3}$$
$$\frac{\overset{1}{\cancel{3}}x}{\underset{1}{\cancel{3}}} = \frac{9}{3}$$
$$x = 3$$

To check, replace *x* with 3:

$$3x + 4 + 2 = 12 + 3$$
$$3(3) + 4 + 2 \stackrel{?}{=} 12 + 3$$
$$9 + 4 + 2 \stackrel{?}{=} 12 + 3$$
$$13 + 2 \stackrel{?}{=} 15$$
$$15 = 15 \checkmark$$

Example 16: Solve for *x*.

$$4x + 2x + 4 = 5x + 3 + 11$$

Simplify each side.

$$6x + 4 = 5x + 14$$

Subtract 4 from both sides.

$$\begin{array}{rcl} 6x + 4 & = & 5x + 14 \\ \underline{\quad -4} & & \underline{\quad\quad -4} \\ 6x \quad & = & 5x + 10 \end{array}$$

Subtract 5*x* from both sides.

$$\begin{array}{rcl} 6x & = & 5x + 10 \\ \underline{-5x} & & \underline{-5x \quad\quad} \\ x & = & \quad\quad 10 \end{array}$$

To check, replace x with 10:

$$4x + 2x + 4 = 5x + 3 + 11$$

$$4(10) + 2(10) + 4 \stackrel{?}{=} 5(10) + 3 + 11$$

$$40 + 20 + 4 \stackrel{?}{=} 50 + 3 + 11$$

$$60 + 4 \stackrel{?}{=} 53 + 11$$

$$64 = 64 \checkmark$$

Chapter Check-Out

1. Give the algebraic expression for six less than the product of two and n.
2. Evaluate $5a + 3b + c$ if $a = 2$, $b = 3$, and $c = 4$.
3. Evaluate $-\frac{2x}{5y} + \frac{3x}{y}$ if $x = 5$ and $y = -2$.
4. Solve for x: $2x - 7 = 27$.
5. Solve for x: $\frac{3}{4}x + 12 = 20$.
6. Solve for x: $5x + 3 = 2x - 15$.

Answers: 1. $2n - 6$ **2.** 23 **3.** $-6\frac{1}{2}$ **4.** $x = 17$ **5.** $x = 10\frac{2}{3}$ **6.** $x = -6$

Chapter 14
WORD PROBLEMS

Chapter Check-In

- ❑ Solving word problems
- ❑ Key words
- ❑ Average problems
- ❑ Money problems
- ❑ Percent problems
- ❑ Distance problems
- ❑ Proportion problems
- ❑ Measurement problems

Mathematical word problems often bring needless fear and anxiety to math students. Don't let the descriptive words surrounding the important numbers and information scare you or mislead you. The following solving process can help you simplify what appears to be a difficult word problem.

Solving Process

A solving process is a step-by-step method to assist you in approaching word problems in an organized, focused, and systematic manner.

Step 1: Find and underline or circle what the question is asking. Identify what you are trying to find.

How <u>tall</u> is the girl? What is the <u>cost</u>? How <u>fast</u> is the car? Underlining or circling what you are looking for helps you make sure that you are answering the question.

Step 2: Focus on and pull out important information in the problem.

Watch for key words that help give you a relationship between the values given.

Step 3: Set up the work that is needed—that is, the operations necessary to answer the question.

This may be setting up a basic operation such as multiplication, setting up a ratio or proportion, or setting up an equation.

Step 4: Do the necessary work or computation carefully.

One of the most common and annoying mistakes is to set up the problem correctly and then make a simple computational error.

Step 5: Put your answer into a sentence to make sure that you answered the question being asked.

Another common error is the failure to answer what was being asked.

Step 6: Check to make sure that your answer is reasonable.

A simple computational error, such as accidentally adding a zero, can give you a ridiculous answer. Estimating an answer can often save you from this type of mistake.

Key Words

The following key words help you understand the relationships between the pieces of information given and give you clues as to how the problem should be solved.

Add

- **Addition:** as in *The team needed the addition of three new players. . .*
- **Sum:** as in *The sum of 5, 6, and 8. . .*
- **Total:** as in *The total of the last two games. . .*
- **Plus:** as in *Three chairs plus five chairs. . .*
- **Increase:** as in *Her pay was increased by $30. . .*

Subtract

- **Difference:** as in *What is the difference between 8 and 5. . .*
- **Fewer:** as in *There were ten fewer girls than boys. . .*

- **Remainder:** as in *What is the remainder when. . .* or *How many are left when. . .*
- **Less:** as in *A number is six less than another number. . .*
- **Reduced:** as in *His allowance was reduced by $5. . .*
- **Decreased:** as in *What number decreased by 7 is 5. . .*
- **Minus:** as in *Seven minus some number is. . .*

Multiply

- **Product:** as in *The product of 3 and 6 is . . .*
- **Of:** as in *One-half of the people in the room. . .*
- **Times:** as in *Six times as many men as women. . .*
- **At:** as in *The cost of five yards of material at $9 a yard is. . .*
- **Total:** As in *If you spend $20 per week on gas, what is the total for a two-week period. . .*
- **Twice:** as in *Twice the value of some number. . .* (multiplying by 2)

Divide

- **Quotient:** as in *The final quotient is. . .*
- **Divided by:** as in *Some number divided by 5 is. . .*
- **Divided into:** as in *The coins were divided into groups of. . .*
- **Ratio:** as in *What is the ratio of. . .*
- **Half:** as in *Half of the cards were. . .*(dividing by 2)

As you practice working word problems, you will discover more key words and phrases that give you insight into the solving process.

Example 1: Jack bowled four games for a total score of 500. What was his average score for a game?

Step 1: Find and underline or circle what the question is asking.

What was <u>his average score for a game</u>?

Step 2: Focus on and pull out important information.

four games for a total score of 500

Step 3: Set up the work that is needed.

$$500 \div 4$$

(The total divided by the number of games gives the average.)

Step 4: Do the necessary work or computation carefully.

$$4\overline{)500}\quad = 125$$

Step 5: Put your answer into a sentence to make sure that you answered the question being asked.

Jack's average score for a game is 125.

Step 6: Check to make sure that your answer is reasonable.

Since four games of 125 total 500, your answer is reasonable and correct.

Example 2: Judy scored 85, 90, and 95 on her last three algebra tests. What was her average score for these tests?

Step 1: Find and underline or circle what the question is asking.

What was <u>her average score for these tests</u>?

Step 2: Focus on and pull out important information.

85, 90, and 95 on three tests

Step 3: Set up the work that is needed.

$$(85 + 90 + 95) \div 3 =$$

(The total divided by the number of scores gives the average.)

Step 4: Do the necessary work or computation carefully.

$$\frac{(85+90+95)}{3} = \frac{270}{3}$$
$$= 90$$

Step 5: Put your answer into a sentence to make sure that you answered the question being asked.

Judy's average test score for these tests was 90.

Step 6: Check to make sure that your answer is reasonable.

Since her scores were 85, 90, and 95, the average should be halfway between 85 and 95. So 90 is a reasonable answer.

Example 3: Frances goes to the market and buys two boxes of cereal at \$4 each, three bottles of milk at \$2 each, and two cans of soup at \$1 each. How much change will Frances get from a \$20 bill?

Step 1: Find and underline or circle what the question is asking.

How much change will Frances get from a \$20 bill?

Step 2: Focus on and pull out important information.

two boxes at \$4 each

three bottles at \$2 each

two cans at \$1 each

\$20 bill used

Step 3: Set up the work that is needed.

$$2 \times 4 =$$

$$3 \times 2 =$$

$$2 \times 1 =$$

$$\$20 - ? =$$

Step 4: Do the necessary work or computation carefully.

$$2 \times 4 = 8$$

$$3 \times 2 = 6$$

$$2 \times 1 = 2$$

$$8 + 6 + 2 = 16$$

$$20 - 16 = 4$$

Step 5: Put your answer into a sentence to make sure that you answered the question being asked.

Frances will get \$4 change.

Step 6: Check to make sure that your answer is reasonable.

Since the total expenses were \$16, then \$4 change from a \$20 bill is reasonable.

Example 4: Sarah can purchase a television for $275 cash or for $100 as a down payment and ten monthly payments of $30 each. How much money can Sarah save by paying cash for the television?

Step 1: Find and underline or circle what the question is asking.

How much money can Sarah save by paying cash for the television?

Step 2: Focus on and pull out important information.

cash $275

$100 down plus ten payments of $30

Step 3: Set up the work that is needed.

$$100 + (10 \times 30) =$$

$$? - 275 =$$

Step 4: Do the necessary work or computation carefully.

$$100 + (10 \times 30) = 100 + 300$$

$$= 400$$

$$400 - 275 = 125$$

Step 5: Put your answer into a sentence to make sure that you answered the question being asked.

Sarah can save $125 by paying cash.

Step 6: Check to make sure that your answer is reasonable.

Ten payments of $30 each is $300 plus a $100 down payment gives $400. This is $125 more than $275, so the answer is reasonable.

Example 5: If apples sell for $3.25 per dozen, how many apples can Maria buy for $13.00?

Step 1: Find and underline or circle what the question is asking.

How many apples can Maria buy for $13.00?

Step 2: Focus on and pull out important information.

$3.25 per dozen

$13.00

Step 3: Set up the work that is needed.

$$\$13.00 \div 3.25 = ?\text{ dozen}$$

$$= ?\text{ apples}$$

Step 4: Do the necessary work or computation carefully.

$$\$13.00 \div 3.25 = 4 \text{ dozen}$$
$$= 4 \times 12$$
$$= 48$$

Step 5: Put your answer into a sentence to make sure that you answered the question being asked.

Maria can buy 48 apples for $13.00.

Step 6: Check to make sure that your answer is reasonable.

Since Maria could buy 12 apples for about $3, then it is reasonable that she could buy 48 apples for about $13.

Example 6: Sequoia Junior High School has a student enrollment of 2,000. If 30% of the students are seventh graders, how many seventh graders are enrolled at the school?

Step 1: Find and underline or circle what the question is asking.

How many seventh graders are enrolled at the school?

Step 2: Focus on and pull out important information.

2,000 students

30% are seventh graders

Step 3: Set up the work that is needed.

$$30\% \text{ of } 2{,}000 =$$

Step 4: Do the necessary work or computation carefully.

$$30\% \text{ of } 2{,}000 = 0.30 \times 2{,}000$$
$$= 600$$

Step 5: Put your answer into a sentence to make sure that you answered the question being asked.

There are 600 seventh graders enrolled at Sequoia Junior High.

Step 6: Check to make sure that your answer is reasonable.

Since 30% of 1,000 is 300, then 30% of 2,000 is 600. The answer is reasonable.

Example 7: Jim Chamberlain, the center for the West Hills Basketball Stars, makes 75% of his free throws. If Jim attempts 80 free throws in a season, how many of his free throws does he make?

Step 1: Find and underline or circle what the question is asking.

How many of his free throws does he make?

Step 2: Focus on and pull out important information.

makes 75%

shoots 80 free throws

Step 3: Set up the work that is needed.

$$75\% \text{ of } 80 =$$

Step 4: Do the necessary work or computation carefully.

$$\begin{aligned} 75\% \text{ of } 80 &= 0.75 \times 80 \\ &= 60 \end{aligned}$$

or

$$\begin{aligned} 75\% \text{ of } 80 &= \frac{3}{\cancel{4}_1} \times \frac{\cancel{80}^{20}}{1} \\ &= 60 \end{aligned}$$

Step 5: Put your answer into a sentence to make sure that you answered the question being asked.

Jim makes 60 free throws.

Step 6: Check to make sure that your answer is reasonable.

Since 50% of his 80 free throws would be 40, then it's reasonable for 75% to be 60.

Example 8: Each week, John spends $50 of his income on entertainment. If John earns $200 a week, what percent of his income is spent on entertainment?

Step 1: Find and underline or circle what the question is asking.

What percent of his income is spent on entertainment?

Step 2: Focus on and pull out important information.

$50 on entertainment

$200 income

Step 3: Set up the work that is needed.

$$\frac{50}{200} = \frac{\text{entertainment}}{\text{income}}$$

Step 4: Do the necessary work or computation carefully.

$$\frac{50}{200} = \frac{1}{4}$$
$$= 25\%$$

Step 5: Put your answer into a sentence to make sure that you answered the question being asked.

John spends 25% of his income on entertainment.

Step 6: Check to make sure that your answer is reasonable.

Since 50 is half, or 50% of 100, then 50 would reasonably be 25% of 200.

Example 9: The Gomez family spends 30% of its income for food. If the family spent $6,000 for food last year, what was the family income for last year?

Step 1: Find and underline or circle what the question is asking.

What was the <u>family income for last year</u>?

Step 2: Focus on and pull out important information.

30% of income for food

$6,000 spent for food last year

Step 3: Set up the work that is needed.

$$30\% \text{ of income} = 6{,}000$$

$$30\%x = 6{,}000$$

Step 4: Do the necessary work or computation carefully.

$$30\%x = 6{,}000$$

$$0.30x = 6{,}000$$

$$\frac{0.30x}{0.30} = \frac{6{,}000}{0.30}$$

$$\frac{\overset{1}{\cancel{0.30}}\,x}{\underset{1}{\cancel{0.30}}} = \frac{6{,}000}{0.30}$$

$$x = 20{,}000$$

Step 5: Put your answer into a sentence to make sure that you answered the question being asked.

The Gomez family income for last year was $20,000.

Step 6: Check to make sure that your answer is reasonable.

Since 30% of the family income is spent for food, and 30% of $20,000 is $6,000, then the answer is reasonable.

Example 10: A miniature piano keyboard is $16\frac{1}{2}$ inches wide. If each key is $1\frac{1}{2}$ inches wide, how many keys are there?

Step 1: Find and underline or circle what the question is asking.

How many keys are there?

Step 2: Focus on and pull out important information.

$16\frac{1}{2}$ inches wide

$1\frac{1}{2}$ inch keys

Step 3: Set up the work that is needed.

$16\frac{1}{2} \div 1\frac{1}{2} =$ or $16.5 \div 1.5 =$

Step 4: Do the necessary work or computation carefully.

$$\begin{aligned} 16\frac{1}{2} \div 1\frac{1}{2} &= \frac{33}{2} \div \frac{3}{2} \\ &= \frac{33}{2} \times \frac{2}{3} \\ &= \frac{\overset{11}{\cancel{33}}}{\underset{1}{\cancel{2}}} \times \frac{\overset{1}{\cancel{2}}}{\underset{1}{\cancel{3}}} \\ &= 11 \end{aligned}$$

or

$$\begin{aligned} 16.5 \div 1.5 &= \frac{16.5}{1.5} \\ &= \frac{165}{15} \\ &= 11 \end{aligned}$$

Step 5: Put your answer into a sentence to make sure that you answered the question being asked.

There are 11 keys on the miniature keyboard.

Step 6: Check to make sure that your answer is reasonable.

Since 16 divided by 1 is 16, and 16 divided by 2 is 8, the answer of 11 is reasonable.

Example 11: The low temperature on Big Bear mountain was 30 degrees on Monday, 20 degrees on Tuesday, -10 degrees on Wednesday, and 15 degrees on Thursday. If you total the changes in low temperature from each day to the next, what is the total number of degrees change?

Step 1: Find and underline or circle what the question is asking.

What is the total number of degrees change?

Step 2: Focus on and pull out important information.

30 on Monday

20 on Tuesday

–10 on Wednesday

15 on Thursday

Step 3: Set up the work that is needed.

$$30 \text{ to } 20 =$$

$$20 \text{ to } -10 =$$

$$-10 \text{ to } 15 =$$

Step 4: Do the necessary work or computation carefully.

$$30 \text{ to } 20 = -10$$

$$20 \text{ to } -10 = -30$$

$$-10 \text{ to } 15 = 25$$

$$\text{total change} = -10 + -30 + 25$$

$$= -15$$

Step 5: Put your answer into a sentence to make sure that you answered the question being asked.

The low temperature changed –15 degrees during the days Monday through Thursday.

Step 6: Check to make sure that your answer is reasonable.

Since the low temperature dropped 10, dropped 30, and rose 25, the total of –15 is correct and reasonable.

Example 12: The Silverado Flash, a solar-powered land vehicle, travels at a maximum speed of 97 miles per hour. At this rate, how far will the Silverado Flash travel in 15 hours?

An important formula to remember is $d = rt$, or distance equals rate times time.

Step 1: Find and underline or circle what the question is asking.

How far will the Silverado Flash travel in 15 hours?

Step 2: Focus on and pull out important information.

97 miles per hour

15 hours

Step 3: Set up the work that is needed.

$$\text{distance} = \text{rate} \times \text{time}$$

$$d = 97 \times 15$$

Step 4: Do the necessary work or computation carefully.

$$d = 97 \times 15$$

$$= 1{,}455$$

Step 5: Put your answer into a sentence to make sure that you answered the question being asked.

The Silverado Flash will travel 1,455 miles in 15 hours.

Step 6: Check to make sure that your answer is reasonable.

At 100 miles per hour for 15 hours, the Silverado Flash would have traveled 1,500 miles. So the answer of 1,455 miles is reasonable.

Example 13: Asaf can run around the track, 440 yards, in 65 seconds. At this same rate, how far could Asaf run in 195 seconds?

Step 1: Find and underline or circle what the question is asking.

How far could Asaf run in 195 seconds?

Step 2: Focus on and pull out important information.

440 yards in 65 seconds

195 seconds

Step 3: Set up the work that is needed.

You could set up the proportion

$$\frac{440 \text{ yards}}{65 \text{ seconds}} = \frac{x \text{ yards}}{195 \text{ seconds}}$$

or simply divide 195 by 65 and multiply that result by 440.

$$(195 \div 65) \times 440 = ?$$

Step 4: Do the necessary work or computation carefully.

$$\frac{440}{65} = \frac{x}{195}$$

$$65x = (440)(195)$$

$$\frac{\overset{1}{\cancel{65}}x}{\underset{1}{\cancel{65}}} = \frac{(440)(\overset{3}{\cancel{195}})}{\underset{1}{\cancel{65}}}$$

$$x = 1{,}320$$

Step 5: Put your answer into a sentence to make sure that you answered the question being asked.

Asaf could run 1,320 yards in 195 seconds.

Step 6: Check to make sure that your answer is reasonable.

If Asaf ran 400 yards every 60 seconds, or one minute, and since 195 seconds is just over three minutes, then 3 times 400, or 1,200, would be a good approximation. So 1,320 yards is a reasonable answer.

Example 14: A researcher tagged 100 frogs in a nearby pond. One week later she took a sample and only 5 out of 20 frogs were tagged. Using this method, how many frogs would she approximate are in the pond?

Step 1: Find and underline or circle what the question is asking.

How many frogs would she approximate are in the pond?

Step 2: Focus on and pull out important information.

100 tagged

5 out of 20 tagged

Step 3: Set up the work that is needed.

You could set up the proportion.

$$\frac{5\text{ tagged}}{20\text{ total}} = \frac{100\text{ tagged}}{x}$$

Step 4: Do the necessary work or computation carefully.

$$\frac{5}{20} = \frac{100}{x}$$

$$5x = 2{,}000$$

$$\frac{\overset{1}{\cancel{5}}x}{\underset{1}{\cancel{5}}} = \frac{2{,}000}{5}$$

$$x = 400$$

Step 5: Put your answer into a sentence to make sure that you answered the question being asked.

She would approximate that there are 400 frogs in the pond.

Step 6: Check to make sure that your answer is reasonable.

Since she tagged 100, and 5 out of 20, or 1 out of 4, came out tagged, then it is reasonable to have a total of 400 frogs in the pond.

Example 15: Julia has $300 in the bank. She works at a bakery and makes $40 per day. If she deposits all of her earnings in the bank and does not make any withdrawals, how many days of work will it take for her to have $740 in the bank?

Step 1: Find and underline or circle what the question is asking.

How many days of work will it take for her to have $740 in the bank?

Step 2: Focus on and pull out important information.

$300 in the bank

$40 per day

$740 total

Step 3: Set up the work that is needed.

Let d stand for the number of days. Then $40d$ is the amount of money earned in d days, and $40d + 300$ is the amount of money she would have in the bank at the end of d days. So

$$40d + 300 = 740$$

Step 4: Do the necessary work or computation carefully.

$$\begin{aligned} 40d + 300 &= 740 \\ -300 & \quad -300 \\ \hline 40d &= 440 \end{aligned}$$

$$\frac{\overset{1}{\cancel{40}}\,d}{\underset{1}{\cancel{40}}} = \frac{440}{40}$$

$$d = 11$$

Step 5: Put your answer into a sentence to make sure that you answered the question being asked.

Julia would have to work 11 days to have $740 in the bank.

Step 6: Check to make sure that your answer is reasonable.

At $40 per day, if she worked 11 days, she would have $440. Add this to the $300 she already had, and the total of $740 is correct and reasonable.

Example 16: A train is 50 miles from Seattle. It is traveling away from Seattle at a speed of 60 miles per hour. In how many hours will the train be 290 miles from Seattle?

Step 1: Find and underline or circle what the question is asking.

In <u>how many hours</u> will the train be <u>290 miles from Seattle</u>?

Step 2: Focus on and pull out important information.

50 miles from Seattle

60 miles per hour

290 miles

Step 3: Set up the work that is needed.

Let h stand for the number of hours. Then $60h$ is the distance traveled in h hours, and $60h + 50$ is the distance from Seattle in h hours. So

$$60h + 50 = 290$$

Step 4: Do the necessary work or computation carefully.

$$\begin{aligned} 60h + 50 &= 290 \\ -50 \quad & \quad -50 \\ \hline 60h \quad &= 240 \end{aligned}$$

$$\frac{60h}{60} = \frac{240}{60}$$

$$\frac{\overset{1}{\cancel{60}}\,h}{\underset{1}{\cancel{60}}} = \frac{240}{60}$$

$$h = 4$$

Step 5: Put your answer into a sentence to make sure that you answered the question being asked.

It will take the train four hours to be 290 miles from Seattle.

Step 6: Check to make sure that your answer is reasonable.

At 60 miles per hour, in four hours the train will travel 240 miles. since it is starting 50 miles from Seattle, the total would be 290 miles.

Example 17: A number n is increased by 25, and the outcome is 57. What is the value of the number n?

Step 1: Find and underline or circle what the question is asking.

What is the <u>value</u> of the number <u>n</u>?

Step 2: Focus on and pull out important information.

n is increased by 25

outcome is 57

Step 3: Set up the work that is needed.

$$n + 25 = 57$$

Step 4: Do the necessary work or computation carefully.

$$\begin{aligned} n + 25 &= 57 \\ \underline{-25} \quad & \underline{-25} \\ n \quad &= 32 \end{aligned}$$

Step 5: Put your answer into a sentence to make sure that you answered the question being asked.

The value of the number is 32.

Step 6: Check to make sure that your answer is reasonable.

Since 32 plus 25 equals 57, the answer checks and is reasonable.

Example 18: The length of a Lambo Speed Wagon is 75 centimeters less than the length of a Corvette. The Lambo is 410 centimeters long. How long is a Corvette?

Step 1: Find and underline or circle what the question is asking.

<u>How long</u> is a <u>Corvette</u>?

Step 2: Focus on and pull out important information.

Lambo is 75 centimeters less

Lambo is 410 centimeters

Step 3: Set up the work that is needed.

Let c represent the length of a Corvette. Then $c - 75$ is the length of a Lambo. So

$$c - 75 = 410$$

Step 4: Do the necessary work or computation carefully.

$$\begin{aligned} c - 75 &= 410 \\ \underline{+75} \quad & \underline{+75} \\ c \quad &= 485 \end{aligned}$$

Step 5: Put your answer into a sentence to make sure that you answered the question being asked.

The length of a Corvette is 485 centimeters.

Step 6: Check to make sure that your answer is reasonable.

Since 485 - 75 is 410, the answer checks and is reasonable.

Chapter Check-Out

1. If it took Silvia five days to save $75, how much did she average per day?
2. Alder Middle School has a total enrollment of 1,200 students. If 25% of the students are seniors, how many seniors are enrolled at the school?
3. Shalla spends 20% of her monthly allowance on clothes. If Shalla spent $15 on clothes last month, what is her monthly allowance?
4. Sam walks at a rate of 3 miles per hour. At this rate, how far can he walk in 6.5 hours?
5. Rick was trying to predict how many students would vote for Hal in the school presidential election. Of the 50 students he randomly selected, 31 said they were voting for Hal. If there are 2,000 students in the school, approximately how many students can he estimate will vote for Hal?
6. Shelly lives 30 miles east of Burbank. If she drives east at a rate of 65 miles per hour, how long will it take her to be 225 miles east of Burbank?
7. A number x is decreased by 15, and the outcome is 70. What is the value of the number x?
8. Felix cuts a board that is 50 inches long into eight equal pieces. How long is each piece?

Answers: 1. $15 **2.** 300 students **3.** $75 **4.** $19\frac{1}{2}$ miles **5.** 1,240 students **6.** 3 hours **7.** 85 **8.** 6.25 inches or $6\frac{1}{4}$ inches

REVIEW QUESTIONS

Chapter 1

1. Which of the following are integers? 2.5, –8, 0, 5
2. Which of the following are prime numbers? 2, 5, 6, 7, 15
3. True or False: An example of the commutative property of addition is $9 + 7 = 7 + 9$.
4. What is the multiplicative inverse of $\frac{1}{4}$?
5. Simplify $3[4 + 6(2 + 6) + 3]$.

Chapter 2

6. Write 7,123 in expanded notation.
7. Round off 327,368 to the nearest thousand.
8. Estimate the product of 269×301 by rounding to the nearest hundred.
9. 6,732 is divisible by which of the following numbers? 2, 3, 4, 5, 6, 7, 8, 9
10. Express 50 as a product of prime factors.

Chapter 3

11. Write 6.12 in expanded notation.
12. Which is greater, 0.66 or 0.658?
13. Round off 0.4445 to the nearest hundredth.
14. Add $27.44 + 3.9 + 100.001$.
15. Multiply 303.2×0.03.
16. Estimate the product of 6.4×4.9 by rounding to the nearest whole number.

Chapter 4

17. Change $\frac{8}{3}$ into a mixed number.

18. Change $6\frac{1}{3}$ into an improper fraction.

19. Is $\frac{3}{5}$ equivalent to $\frac{5}{8}$?

20. Add $5\frac{1}{5}+3\frac{1}{4}$.

21. Subtract $8\frac{1}{6}-4\frac{3}{4}$.

22. Multiply $7\frac{2}{3}\times 3\frac{1}{2}$.

23. Divide $5\frac{1}{6}\div 3\frac{2}{5}$.

24. Simplify $2+\dfrac{2}{2+\frac{1}{2}}$.

25. Change $\frac{1}{8}$ to a decimal.

26. Change 0.08 to a fraction in lowest terms.

27. Change $0.111...$ to a fraction.

Chapter 5

28. Change $\frac{6}{25}$ to a percent.

29. Change 40% to a fraction in lowest terms.

30. What is 30% of 80?

31. 24 is what percent of 300?

32. 50 is 20% of what number?

33. What is the percent change from 45 to 60?

Chapter 6

34. $-17 + 9 =$ ____.

35. $-15 - 18 =$ ____.

36. $46 - (-5) =$ ____.

37. $19 - (+6 - 7 + 8 - 5) =$ ____.

38. $(-17) \times (-12) =$ ____.

39. $|-6| =$ ____.

40. $-\frac{1}{5} + \frac{2}{7} =$ ____.

41. $-\frac{3}{5} - \frac{7}{8} =$ ____.

42. $-4\frac{5}{6} \times 3\frac{1}{2} =$ ____.

43. $\frac{8}{9} \div \frac{6}{7} =$ ____.

Chapter 7

44. $8^{-2} =$ ____.

45. $5^2 \times 5^8 =$ ____. (with exponents)

46. $(3^4)^2 =$ ____. (with exponents)

47. $\sqrt[3]{64} =$ ____.

48. Approximate $\sqrt{83}$ to the nearest tenth.

49. Simplify $\sqrt{40}$.

Chapter 8

50. Multiply 0.01×0.001 and leave the answer in powers of ten.

51. Divide $1{,}000 \div 100$ and leave the answer in powers of ten.

52 Express 45,000 in scientific notation.

53. $(5 \times 10^4)(3 \times 10^3) =$ ____. (in scientific notation)

54. $(8 \times 10^4) \div (4 \times 10^{-2}) =$ ____. (in scientific notation)

Chapter 9

55. How many yards are in 1 mile?

56. If a kilometer equals 1,000 meters, and 1 dekameter equals 10 meters, how many dekameters are in 8 kilometers?

57. Which of the following is more precise, 7.4 mm or 4.57 mm?

58. Find the number of units used and the significant digits for 23.4 m (unit of measure, 0.1 m).

59. What are the perimeter and area of the triangle?

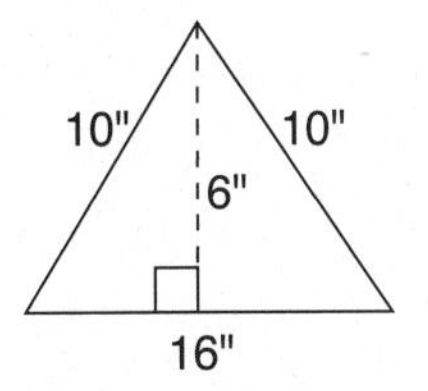

60. What are the perimeter and area of the parallelogram?

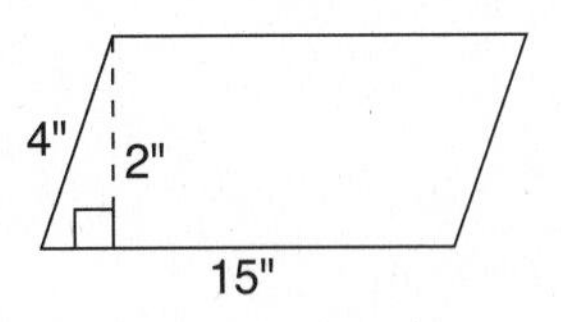

61. What are the perimeter and area of the trapezoid?

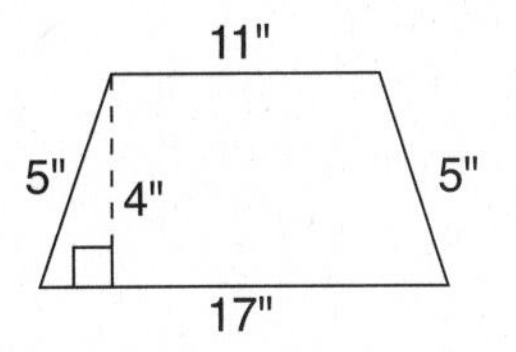

62. What are the circumference and area of the circle? (Leave answer in terms of π.)

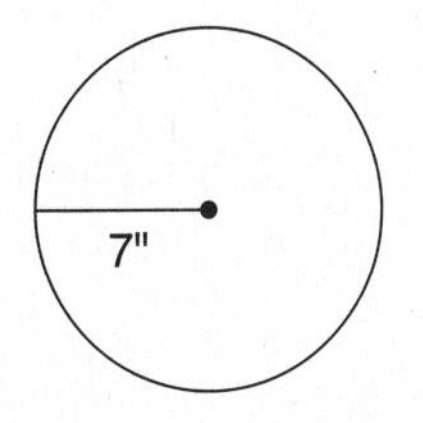

Chapter 10

63. The following bar graph indicates that Bob saved approximately how much more money than Sandra?

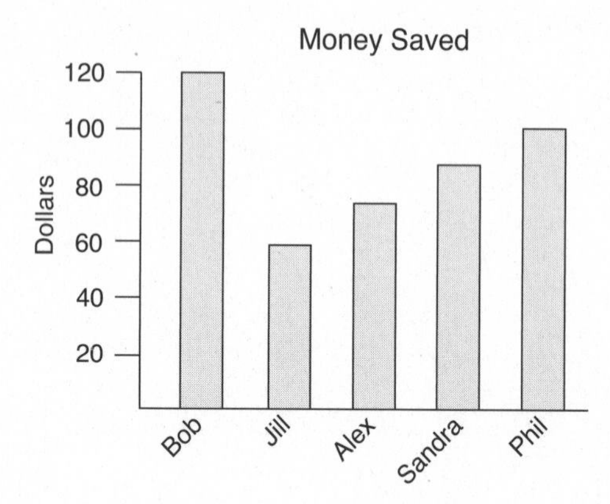

64. According to the line graph, Elisa increased her driving speed the most between which two streets?

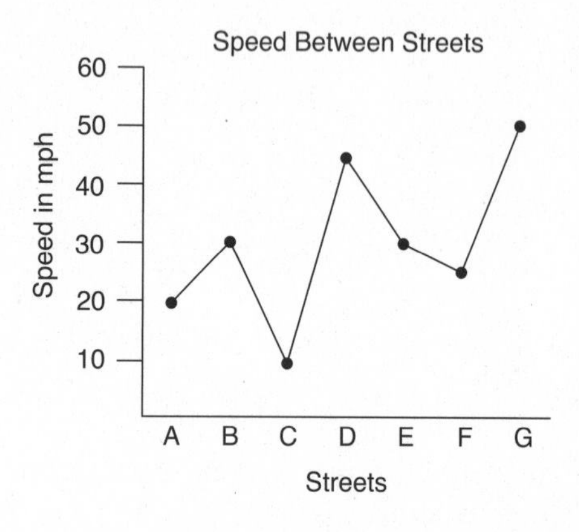

65. Based on the circle graph, how much money does Sammy spend on food?

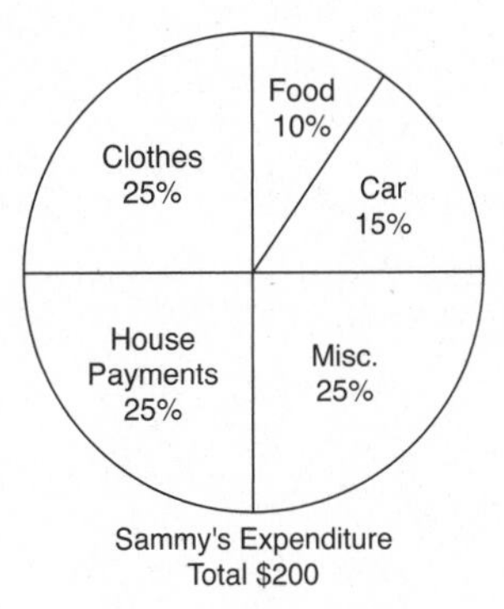

66. Identify the coordinates of the points *A* and *B* on the coordinate graph.

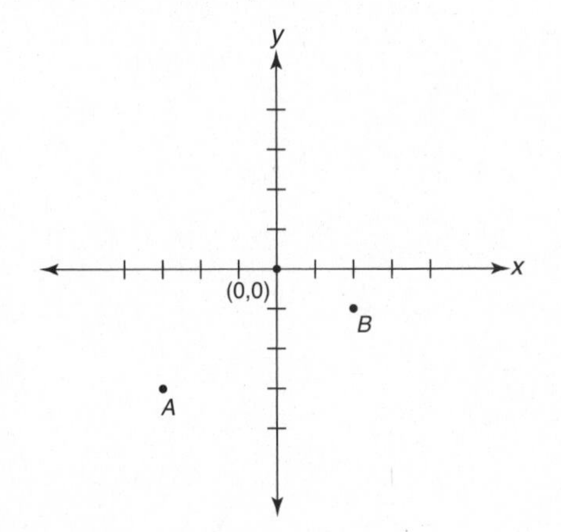

Chapter 11

67. Using the spinner with equal sections shown here, what is the probability of spinning either a 2 or an 8 in one spin?

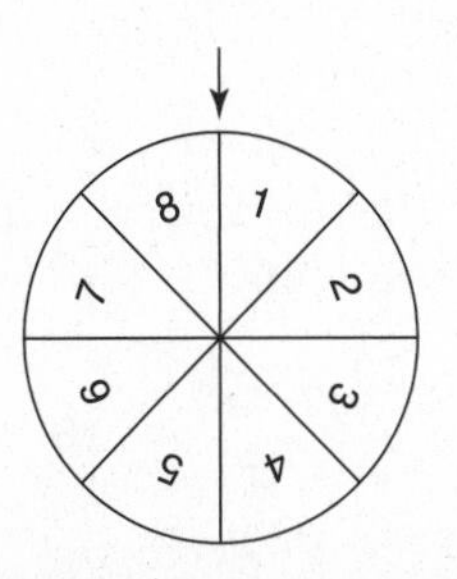

68. What is the probability of rolling two dice in one toss so they total 9?

69. How many possible combinations of jackets, shirts, and pants, are there if there are four different-color jackets, three different-color shirts, and two different-color pants?

70. How many ways can you arrange the letters T, E, A, M, S in a row?

71. What is the arithmetic mean, mode, and median of the following quiz scores: 6, 7, 7, 7, 8, 4, 4, 6, 3, 7, 1, 0?

Chapter 12

72. What is the next number in the progression 5, 9, 13, 17, ... ?

73. What is the fiftieth number in the progression 2, 4, 6, 8, ... ?

74. What is the next number in the progression 4, 16, 64, 256, ... ?

Chapter 13

75. Give the algebraic expression for five more than the product of three and n.

76. Evaluate: $5a + 3b + c$ if $a = 2$, $b = 3$, and $c = 4$.

77. Solve for x: $2x + 7 = 27$.

78. Solve for x: $5x + 3 = 2x - 15$.

Chapter 14

79. Rajiv bowled four games with scores of 170, 160, 150, and 130. What was his average score?

80. Steve goes to the bookstore and buys two books at $5.90 each and one book for $8.00. If sales tax is included in the prices, how much change will Steve get from a $20 bill?

81. Tina pitched 45 strikes in her softball game. If 60% of her total pitches were strikes, how many total pitches did she throw?

82. Two times a number n is increased by 50 giving an outcome of 200. What is the value of the number n?

Answers: **1.** –8, 0, 5 **2.** 2, 5, 7 **3.** True **4.** $\frac{4}{1}$ or 4 **5.** 165 **6.** $(7 \times 1{,}000) + (1 \times 100) + (2 \times 10) + (3 \times 1)$ or $(7 \times 10^3) + (1 \times 10^2) + (2 \times 10^1) + (3 \times 10^0)$ **7.** 327,000 **8.** 90,000 **9.** 2, 3, 4, 6, 9 **10.** $2 \times 5 \times 5$ or 2×5^2 **11.** $(6 \times 1) + (1 \times 0.1) + (2 \times 0.01)$ or $(6 \times 10^0) + (1 \times 10^{-1}) + (2 \times 10^{-2})$ **12.** 0.66 **13.** 0.44 **14.** 131.341 **15.** 9.096 **16.** 30 **17.** $2\frac{2}{3}$ **18.** $\frac{19}{3}$ **19.** No **20.** $8\frac{9}{20}$ **21.** $3\frac{5}{12}$ **22.** $\frac{161}{6}$ or $26\frac{5}{6}$ **23.** $\frac{155}{102}$ or $1\frac{53}{102}$ **24.** $2\frac{4}{5}$ **25.** 0.125 **26.** $\frac{2}{25}$ **27.** $\frac{1}{9}$ **28.** 24% **29.** $\frac{2}{5}$ **30.** 24 **31.** 8% **32.** 250 **33.** $33\frac{1}{3}\%$ **34.** –8 **35.** –33 **36.** 51 **37.** 17 **38.** 204 **39.** 6 **40.** $\frac{3}{35}$ **41.** $-\frac{59}{40}$ or $-1\frac{19}{40}$ **42.** $-\frac{203}{12}$ or $-16\frac{11}{12}$ **43.** $\frac{28}{27}$ or $1\frac{1}{27}$ **44.** $\frac{1}{64}$ **45.** 5^{10} **46.** 3^8 **47.** 4 **48.** 9.1 **49.** $2\sqrt{10}$ **50.** 10^{-5} **51.** 10^1 or 10 **52.** 4.5×10^4 **53.** 1.5×10^8 **54.** 2×10^6 **55.** 1,760 yards **56.** 800 dekameters **57.** 4.57mm **58.** 234 units, significant digits are 2,3,4 **59.** $P = 36$ in, $A = 48$ sq in **60.** $P = 38$ in, $A = 30$ sq in **61.** $P = 38$ in, $A = 56$ sq in **62.** $C = 14\pi$ in, $A = 49\pi$ sq in **63.** \$30 **64.** Streets C and D **65.** \$20 **66.** A(–3,–3) B(2,–1) **67.** $\frac{2}{8}$ or $\frac{1}{4}$ **68.** $\frac{4}{36}$ or $\frac{1}{9}$ **69.** 24 **70.** 120 **71.** Mean = 5, mode = 7, median = 6 **72.** 21 **73.** 100 **74.** 1,024 **75.** $3n + 5$ **76.** 23 **77.** $x = 10$ **78.** $x = -6$ **79.** $152\frac{1}{2}$ **80.** \$0.20 change **81.** 75 total pitches **82.** 75

RESOURCE CENTER

The Resource Center offers the best resources available in print and online to help you study and review the core concepts of basic math and pre-algebra. You can find additional resources, plus study tips and tools to help test your knowledge, at www.cliffsnotes.com.

Books

CliffsNotes Basic Math & Pre-Algebra Quick Review, 2nd Edition, is one of many great books available to help you review, refresh, and relearn mathematics. If you want some additional resources for math review, check out the following publications.

CliffsNotes Math Review for Standardized Tests, 2nd Edition, by Jerry Bobrow, Ph.D., revised by Edward Kohn, M.S., gives you a step-by-step review of arithmetic, algebra, geometry, and word problems. Includes pre-test and post-test for each subject area and glossaries. Published by Wiley $14.99.

CliffsNotes Algebra I Quick Review, 2nd Edition, by Jerry Bobrow, Ph.D., gives you an easy-to-use guide to review, refresh, and relearn Algebra I. Includes chapter check-outs, chapter reviews, and a glossary. Published by Wiley $9.99.

CliffsNotes Geometry Quick Review, 2nd Edition, by Edward Kohn, M.S., gives you an outstanding review of the basic concepts of geometry. Includes chapter check-outs, chapter reviews, and a glossary. Published by Wiley $9.99.

CliffNotes Algebra II Quick Review, 2nd Edition, by Edward Kohn, M.S., and David Alan Herzog, provides an excellent review of Algebra II with clearly explained sample problems. Includes chapter check-outs, chapter reviews, and a glossary. Published by Wiley $9.99.

CliffsNotes Trigonometry Quick Review, 2nd Edition, by David A. Kay, M.S., provides an outstanding review of the basic concepts of trigonometry. Clearly explained example problems and figures make the concepts easier to understand. Includes chapter check-outs, chapter reviews, and a glossary. Published by Wiley $9.99.

Wiley also has three Web sites that you can visit to read about all the books we publish:

- www.cliffsnotes.com
- www.dummies.com
- www.wiley.com

Internet

Visit the following Web sites for more information about basic math and pre-algebra.

Math.com, http://www.math.com/students/homeworkhelp.html. If you can spare a mere 60 seconds, then you can improve your math skills. Math.com's Homework Help offers quick one-minute tips, tricks, and tidbits designed to help you improve your skills in Pre-Algebra, Algebra, and Geometry. Math.com also offers a handy glossary that helps you figure out the difference between a rhombus and a ray.

Discovery.com, http://school.discovery.com/homeworkhelp/webmath/. Have you been working on the same math problem for an hour? Help is on the way! At Discovery.com's WebMath you can find tips and tricks on every type of math problem from Pre-Algebra to Calculus. Just find a math problem similar to the one you're working on, then use WebMath to find out how to solve the problem and check your answer. Your math homework just got a whole lot easier.

Next time you're on the Internet, don't forget to drop by www.cliffsnotes.com. We created an online Resource Center that you can use today, tomorrow, and beyond.

GLOSSARY

additive inverse The opposite (negative) of a number. Any number plus its additive inverse equals 0.

associative property Grouping of elements makes no difference in the outcome; only true for multiplication and addition.

braces Grouping symbols used after the use of brackets; signs { } used to represent a set.

canceling In multiplication of fractions, dividing the same number into both a numerator and a denominator.

circumference The distance around a circle; equals 2 times π times the radius or π times the diameter ($C = 2\pi r$ or πd).

closure property When all answers to the operation performed on a set of numbers fall into the original set.

common denominator A number that can be divided evenly by all denominators in the problem.

common factors Factors that are the same for two or more numbers.

common multiples Multiples that are the same for two or more numbers.

commutative property Order of elements does not make any difference in the outcome; only true for multiplication and addition.

complex fraction A fraction having a fraction or fractions in the numerator and/or denominator.

composite number A number divisible by more than just 1 and itself.

cube The result when a number is multiplied by itself twice.

cube root A number that when multiplied by itself twice gives you the original number; its symbol is $\sqrt[3]{\ }$.

decimal fraction Fraction with a denominator 10, 100, 1,000, and so on, written using a decimal point; for example, .3 and .275.

decimal point A dot/symbol used to distinguish decimal fractions from whole numbers.

denominator The bottom symbol or number of a fraction.

dependent events When the outcome of one event has a bearing or effect on the outcome of another event.

difference The result of subtraction.

distributive property The process of distributing a number on the outside of the parentheses to each number on the inside; $a(b + c) = ab + ac$.

even number An integer divisible by 2.

expanded notation Pointing out the place value of a digit by writing a number as the digit times its place value. For example, $342 = (3 \times 10^2) + (4 \times 10^1) + (2 \times 10^0)$.

exponent A small number placed above and to the right of a number; expresses the power to which the quantity is to be raised or lowered.

factor (noun) A number or symbol that divides evenly into a larger number. For example, 6 is a factor of 24.

factor (verb) To find two or more quantities whose product equals the original quantity.

fraction A symbol that expresses part of a whole and consists of a numerator and a denominator; for example, $\frac{3}{5}$.

greatest common factor The largest factor common to two or more numbers.

identity element for addition is 0. Any number added to 0 gives the original number.

identity element for multiplication is 1. Any number multiplied by 1 gives the original number.

improper fraction A fraction in which the numerator is greater than the denominator; for example, $\frac{3}{2}$.

independent events When the outcome of one event has no bearing or effect on the outcome of another event.

integer A whole number, either positive, negative, or zero.

invert Turn upside down, as in invert $\frac{2}{3}$ to create $\frac{3}{2}$.

irrational number A number that is not rational (cannot be written as a fraction $\frac{x}{y}$); for example, $\sqrt{3}$ or π.

least common multiple The smallest multiple that is common to two or more numbers.

lowest common denominator The smallest number that can be divided evenly by all denominators in the problem.

mean (arithmetic) The average of a number of items in a group (the sum items divided by the number of items).

median The middle item in an ordered group. If the group has an even number of items, the median is the mean of the two middle terms.

mixed number A number containing both a whole number and a fraction; for example, $5\frac{1}{2}$.

mode The number appearing most frequently in a group.

multiples Numbers found by multiplying a number by 2, by 3, by 4, and so on.

multiplicative inverse The reciprocal of a number. Any number multiplied by its multiplicative inverse equals 1.

natural number A counting number; 1, 2, 3, 4, and so on.

negative number A number less than 0.

number sequence A list of numbers with some pattern. One number follows another in some defined manner.

numerator The top symbol or number of a fraction.

odd number An integer not divisible by 2.

operation Multiplication, addition, subtraction, or division.

order of operations The priority given to an operation relative to other operations. For example, multiplication is performed before addition.

percent or percentage A common fraction with 100 as its denominator. For example, 37% is $\frac{37}{100}$.

place value The value given a digit by the position of a digit in the number.

positive number A number greater than zero.

power A product of equal factors. $4 \times 4 \times 4 = 4^3$, reads "four to the third power" or "the third power of four." Power and exponent are sometimes used interchangeably.

prime number A number with exactly two different factors, itself and one.

probability The numerical measure of the likelihood of an outcome or event occurring.

product The result of multiplication.

proper fraction A fraction in which the numerator is less than the denominator; for example, $\frac{2}{3}$.

proportion An equating of two equal ratios. For example, 5 is to 4 as 10 is to 8, or $\frac{5}{4} = \frac{10}{8}$.

quotient The result of division.

range The difference between the largest and the smallest number in a set of numbers.

ratio A comparison between two numbers or symbols; may be written $x{:}y$, x/y, or x is to y.

rational number An integer or fraction such as $\frac{7}{8}$ or $\frac{9}{4}$ or $\frac{5}{1}$. Any number that can be written as a fraction $\frac{x}{y}$ with x being an integer and y a natural number.

real number Any rational or irrational number.

reciprocal The multiplicative inverse of a number. For example, $\frac{2}{3}$ is the reciprocal of $\frac{3}{2}$.

rounding off Changing a number to the nearest place value as specified; a method of approximating.

scientific notation Any number expressed as a value between 1 and 10 and multiplied by a power of 10, used for writing very large or very small numbers; for example, 2.5×10^4.

simplifying Expressing an equation or a fraction in its simplest terms. For example, $\frac{2}{4}$ is simplified to $\frac{1}{2}$.

square The result when a number is multiplied by itself.

square root A number that when multiplied by itself gives you the original number; its symbol is $\sqrt{\ }$. For example, 5 is the square root of 25; $\sqrt{25} = 5$.

sum The result of addition.

tenth The first decimal place to the right of the decimal point. For example, .7 is seven-tenths.

weighted mean The mean of a set of numbers that have been weighted (multiplied by their relative importance or times of occurrence).

whole number 0, 1, 2, 3, and so on.

Index

E

F

G

H

I

K

L

M

N

O

P

Q

R

S